**Crosswalk Coach for the
Common Core State Standards**

Mathematics

Grade 4

Crosswalk Coach for the Common Core State Standards, Mathematics, Grade 4
299NA
ISBN-13: 978-0-7836-7848-1

Contributing Writer: Andie Liao
Cover Image: © Ron Hilton/Dreamstime.com

Triumph Learning® 136 Madison Avenue, 7th Floor, New York, NY 10016

Frequently Asked Questions about the Common Core Standards

What are the Common Core State Standards?

The Common Core State Standards for mathematics and English language arts, grades K–12, are a set of shared goals and expectations for the knowledge and skills that will help students succeed. They allow students to understand what is expected of them and to become progressively more proficient in understanding and using mathematics and English language arts. Teachers will be better equipped to know exactly what they must do to help students learn and to establish individualized benchmarks for them.

Will the Common Core State Standards tell teachers how and what to teach?

No. Because the best understanding of what works in the classroom comes from teachers, these standards will establish *what* students need to learn, but they will not dictate *how* teachers should teach. Instead, schools and teachers will decide how best to help students reach the standards.

What will the Common Core State Standards mean for students?

The standards will provide a clear, consistent understanding of what is expected of student learning across the country. Common standards will not prevent different levels of achievement among students, but they will ensure more consistent exposure to materials and learning experiences through curriculum, instruction, teacher preparation, and other supports for student learning. These standards will help give students the knowledge and skills they need to succeed in college and careers.

Do the Common Core State Standards focus on skills and content knowledge?

Yes. The Common Core Standards recognize that both content and skills are important. They require rigorous content and application of knowledge through higher-order thinking skills. The English language arts standards require certain critical content for all students, including classic myths and stories from around the world, America's founding documents, foundational American literature, and Shakespeare. The remaining crucial decisions about content are left to state and local determination. In addition to content coverage, the Common Core State Standards require that students systematically acquire knowledge of literature and other disciplines through reading, writing, speaking, and listening.

In mathematics, the Common Core State Standards lay a solid foundation in whole numbers, addition, subtraction, multiplication, division, fractions, and decimals. Together, these elements support a student's ability to learn and apply more demanding math concepts and procedures.

The Common Core State Standards require that students develop a depth of understanding and ability to apply English language arts and mathematics to novel situations, as college students and employees regularly do.

Will common assessments be developed?

It will be up to the states: some states plan to come together voluntarily to develop a common assessment system. A state-led consortium on assessment would be grounded in the following principles: allowing for comparison across students, schools, districts, states and nations; creating economies of scale; providing information and supporting more effective teaching and learning; and preparing students for college and careers.

Table of Contents

Common Core State Standards Correlation Chart

Common Core State Standard	Grade 4	*Coach* Lesson(s)
	Domain: Operations and Algebraic Thinking	
	Use the four operations with whole numbers to solve problems.	
4.OA.1	Interpret a multiplication equation as a comparison, e.g., interpret $35 = 5 \times 7$ as a statement that 35 is 5 times as many as 7 and 7 times as many as 5. Represent verbal statements of multiplicative comparisons as multiplication equations.	3
4.OA.2	Multiply or divide to solve word problems involving multiplicative comparison, e.g., by using drawings and equations with a symbol for the unknown number to represent the problem, distinguishing multiplicative comparison from additive comparison.	3, 7
4.OA.3	Solve multi-step word problems posed with whole numbers and having whole-number answers using the four operations, including problems in which remainders must be interpreted. Represent these problems using equations with a letter standing for the unknown quantity. Assess the reasonableness of answers using mental computation and estimation strategies including rounding.	3, 4, 6, 7–9, 12, 13, 15, 16
	Gain familiarity with factors and multiples.	
4.OA.4	Find all factor pairs for a whole number in the range 1–100. Recognize that a whole number is a multiple of each of its factors. Determine whether a given whole number in the range 1–100 is a multiple of a given one-digit number. Determine whether a given whole number in the range 1–100 is prime or composite.	11
	Generate and analyze patterns.	
4.OA.5	Generate a number or shape pattern that follows a given rule. Identify apparent features of the pattern that were not explicit in the rule itself. *For example, given the rule "Add 3" and the starting number 1, generate terms in the resulting sequence and observe that the terms appear to alternate between odd and even numbers. Explain informally why the numbers will continue to alternate in this way.*	17
	Domain: Number and Operations in Base Ten	
	Generalize place value understanding for multi-digit whole numbers.	
4.NBT.1	Recognize that in a multi-digit whole number, a digit in one place represents ten times what it represents in the place to its right. For example, recognize that $700 \div 70 = 10$ by applying concepts of place value and division.	1
4.NBT.2	Read and write multi-digit whole numbers using base-ten numerals, number names, and expanded form. Compare two multi-digit numbers based on meanings of the digits in each place, using $>$, $=$, and $<$ symbols to record the results of comparisons.	1, 2

Common Core State Standard	Grade 4	Coach Lesson(s)
Domain: Number and Operations in Base Ten *(continued)*		
Generalize place value understanding for multi-digit whole numbers. *(continued)*		
4.NBT.3	Use place value understanding to round multi-digit whole numbers to any place.	14
Use place value understanding and properties of operations to perform multi-digit arithmetic.		
4.NBT.4	Fluently add and subtract multi-digit whole numbers using the standard algorithm.	12, 13
4.NBT.5	Multiply a whole number of up to four digits by a one-digit whole number, and multiply two two-digit numbers, using strategies based on place value and the properties of operations. Illustrate and explain the calculation by using equations, rectangular arrays, and/or area models.	4–6, 10
4.NBT.6	Find whole-number quotients and remainders with up to four-digit dividends and one-digit divisors, using strategies based on place value, the properties of operations, and/or the relationship between multiplication and division. Illustrate and explain the calculation by using equations, rectangular arrays, and/or area models.	8–10
Domain: Number and Operations—Fractions		
Extend understanding of fraction equivalence and ordering.		
4.NF.1	Explain why a fraction $\frac{a}{b}$ is equivalent to a fraction $\frac{(n \times a)}{(n \times b)}$ by using visual fraction models, with attention to how the number and size of the parts differ even though the two fractions themselves are the same size. Use this principle to recognize and generate equivalent fractions.	18, 19
4.NF.2	Compare two fractions with different numerators and different denominators, e.g., by creating common denominators or numerators, or by comparing to a benchmark fraction such as $\frac{1}{2}$. Recognize that comparisons are valid only when the two fractions refer to the same whole. Record the results of comparisons with symbols $>$, $=$, or $<$, and justify the conclusions, e.g., by using a visual fraction model.	20
Build fractions from unit fractions by applying and extending previous understandings of operations on whole numbers.		
4.NF.3	Understand a fraction $\frac{a}{b}$ with $a > 1$ as a sum of fractions $\frac{1}{b}$.	
4.NF.3.a	Understand addition and subtraction of fractions as joining and separating parts referring to the same whole.	21, 22
4.NF.3.b	Decompose a fraction into a sum of fractions with the same denominator in more than one way, recording each decomposition by an equation. Justify decompositions, e.g., by using a visual fraction model. *Examples:* $\frac{3}{8} = \frac{1}{8} + \frac{1}{8} + \frac{1}{8}$; $\frac{3}{8} = \frac{1}{8} + \frac{2}{8}$; $2\frac{1}{8} = 1 + 1 + \frac{1}{8} = \frac{8}{8} + \frac{8}{8} + \frac{1}{8}$.	21, 23

Common Core State Standard	Grade 4	*Coach* Lesson(s)
Domain: Number and Operations—Fractions *(continued)*		
**Build fractions from unit fractions by applying and extending previous understandings of operations on whole numbers.** *(continued)*		
4.NF.3.c	Add and subtract mixed numbers with like denominators, e.g., by replacing each mixed number with an equivalent fraction, and/or by using properties of operations and the relationship between addition and subtraction.	23
4.NF.3.d	Solve word problems involving addition and subtraction of fractions referring to the same whole and having like denominators, e.g., by using visual fraction models and equations to represent the problem.	21, 22
4.NF.4	Apply and extend previous understandings of multiplication to multiply a fraction by a whole number.	
4.NF.4.a	Understand a fraction $\frac{a}{b}$ as a multiple of $\frac{1}{b}$. *For example, use a visual fraction model to represent $\frac{5}{4}$ as the product $5 \times \left(\frac{1}{4}\right)$, recording the conclusion by the equation $\frac{5}{4} = 5 \times \left(\frac{1}{4}\right)$.*	24
4.NF.4.b	Understand a multiple of $\frac{a}{b}$ as a multiple of $\frac{1}{b}$, and use this understanding to multiply a fraction by a whole number. *For example, use a visual fraction model to express $3 \times \left(\frac{2}{5}\right)$ as $6 \times \left(\frac{1}{5}\right)$, recognizing this product as $\frac{6}{5}$. (In general, $n \times \left(\frac{a}{b}\right) = \frac{(n \times a)}{b}$.)*	24
4.NF.4.c	Solve word problems involving multiplication of a fraction by a whole number, e.g., by using visual fraction models and equations to represent the problem. *For example, if each person at a party will eat $\frac{3}{8}$ of a pound of roast beef, and there will be 5 people at the party, how many pounds of roast beef will be needed? Between what two whole numbers does your answer lie?*	24
**Understand decimal notation for fractions, and compare decimal fractions.**		
4.NF.5	Express a fraction with denominator 10 as an equivalent fraction with denominator 100, and use this technique to add two fractions with respective denominators 10 and 100. *For example, express $\frac{3}{10}$ as $\frac{30}{100}$, and add $\frac{3}{10} + \frac{4}{100} = \frac{34}{100}$.*	26
4.NF.6	Use decimal notation for fractions with denominators 10 or 100. *For example, rewrite 0.62 as $\frac{62}{100}$; describe a length as 0.62 meters; locate 0.62 on a number line diagram.*	25, 26
4.NF.7	Compare two decimals to hundredths by reasoning about their size. Recognize that comparisons are valid only when the two decimals refer to the same whole. Record the results of comparisons with the symbols $>$, $=$, or $<$, and justify the conclusions, e.g., by using a visual model.	27

Common Core State Standard	Grade 4	Coach Lesson(s)
Domain: Measurement and Data		
Solve problems involving measurement and conversion of measurements from a larger unit to a smaller unit.		
4.MD.1	Know relative sizes of measurement units within one system of units, including km, m, cm; kg, g; lb, oz.; l, ml; hr, min, and sec. Within a single system of measurement, express measurements in a larger unit in terms of a smaller unit. Record measurement equivalents in a two-column table. *For example, know that 1 ft is 12 times as long as 1 in. Express the length of a 4 ft snake as 48 in. Generate a conversion table for feet and inches listing the number pairs (1, 12), (2, 24), (3, 36).*	29–32
4.MD.2	Use the four operations to solve word problems involving distances, intervals of time, liquid volumes, masses of objects, and money, including problems involving simple fractions or decimals, and problems that require expressing measurements given in a larger unit in terms of a smaller unit. Represent measurement quantities using diagrams such as number line diagrams that feature a measurement scale.	28–32
4.MD.3	Apply the area and perimeter formulas for rectangles in real world and mathematical problems. *For example, find the width of a rectangular room given the area of the flooring and the length, by viewing the area formula as a multiplication equation with an unknown factor.*	33, 34
Represent and interpret data.		
4.MD.4	Make a line plot to display a data set of measurements in fractions of a unit $\left(\frac{1}{2}, \frac{1}{4}, \frac{1}{8}\right)$. Solve problems involving addition and subtraction of fractions by using information presented in line plots. *For example, from a line plot find and interpret the difference in length between the longest and shortest specimens in an insect collection.*	36
Geometric measurement: understand concepts of angle and measure angles.		
4.MD.5	Recognize angles as geometric shapes that are formed wherever two rays share a common endpoint, and understand concepts of angle measurement:	
4.MD.5.a	An angle is measured with reference to a circle with its center at the common endpoint of the rays, by considering the fraction of the circular arc between the points where the two rays intersect the circle. An angle that turns through $\frac{1}{360}$ of a circle is called a "one-degree angle," and can be used to measure angles.	35
4.MD.5.b	An angle that turns through *n* one-degree angles is said to have an angle measure of *n* degrees.	35
4.MD.6	Measure angles in whole-number degrees using a protractor. Sketch angles of specified measure.	35
4.MD.7	Recognize angle measure as additive. When an angle is decomposed into non-overlapping parts, the angle measure of the whole is the sum of the angle measures of the parts. Solve addition and subtraction problems to find unknown angles on a diagram in real world and mathematical problems, e.g., by using an equation with a symbol for the unknown angle measure.	35

Common Core State Standard	Grade 4	Coach Lesson(s)
Domain: Geometry		
Draw and identify lines and angles, and classify shapes by properties of their lines and angles.		
4.G.1	Draw points, lines, line segments, rays, angles (right, acute, obtuse), and perpendicular and parallel lines. Identify these in two-dimensional figures.	37, 38
4.G.2	Classify two-dimensional figures based on the presence or absence of parallel or perpendicular lines, or the presence or absence of angles of a specified size. Recognize right triangles as a category, and identify right triangles.	38
4.G.3	Recognize a line of symmetry for a two-dimensional figure as a line across the figure such that the figure can be folded along the line into matching parts. Identify line-symmetric figures and draw lines of symmetry.	39

Domain 1

Number and Operations in Base Ten

Domain 1: Diagnostic Assessment for Lessons 1–10

Lesson 1 Read and Write Whole Numbers
4.NBT.1, 4.NBT.2

Lesson 2 Compare and Order Whole Numbers
4.NBT.2

Lesson 3 Multiplication Facts
4.OA.1, 4.OA.2, 4.OA.3

Lesson 4 Multiply Greater Numbers
4.NBT.5, 4.OA.3

Lesson 5 Multiplication Properties
4.NBT.5

Lesson 6 Distributive Property of Multiplication
4.OA.3, 4.NBT.5

Lesson 7 Division Facts
4.OA.2, 4.OA.3

Lesson 8 Divide Greater Numbers
4.NBT.6, 4.OA.3

Lesson 9 Division with Remainders
4.NBT.6, 4.OA.3

Lesson 10 Multiply and Divide by Multiples of 10, 100, and 1,000
4.NBT.5, 4.NBT.6

Domain 1: Cumulative Assessment for Lessons 1–10

Domain 1: Diagnostic Assessment for Lessons 1–10

1. Frankie read that three hundred sixty-seven thousand, five hundred sixty-two people live in his city. Which is another way to write this number?

   A. 300,000 + 60,000 + 7,000 + 500 + 60 + 2

   B. 367,000,562

   C. 3 + 6 + 7 + 5 + 6 + 2

   D. 367,652

2. How many times greater is the 4 in the hundreds place than the 4 in the ones place in this number?

   536,494

   A. 1

   B. 10

   C. 100

   D. 1,000

3. A spring was 8 centimeters long at first. Now it is stretched to be 72 centimeters long. How many times as long is the spring now as it was at first?

   A. 6

   B. 7

   C. 8

   D. 9

4. Each CD that Rashaun bought last year cost $16. He bought 42 CDs last year. How much money did Rashaun spend on CDs last year?

   A. $672

   B. $472

   C. $462

   D. $96

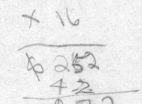

5. What is the missing number in this sentence?

   $$4 \times (5 \times 12) = (\square \times 5) \times 12$$

   A. 1

   B. 4

   C. 5

   D. 12

6. Which number sentence is the same as this sentence?

   $$15 \times 5 = \square$$

   A. (10 + 5) × (5 + 5)

   B. (10 × 5) + (5 × 5)

   C. (10 × 5) × (5 × 5)

   D. (10 + 5) + (5 + 5)

7. Anna spent $36 on 9 notebooks. Each notebook cost the same amount of money. How much did each notebook cost?

   A. $3

   B. $4

   C. $6

   D. $7

8. On Saturday, David read 96 pages of his book. That is 8 times the number of pages he read on Friday. How many pages of his book did he read on Friday?

   A. 8

   B. 9

   C. 12

   D. 14

9. Multiply.

   $3 \times 800 = \boxed{}$

   2,400

   (handwritten: 800 + 800 = 800; 24,00; 2,400)

10. Mr. King has 155 oranges to pack into boxes. Each box can hold 8 oranges. He will keep any extra oranges for himself. How many full boxes of oranges does Mr. King pack?

    A. Write a number sentence for the problem. Use a variable to represent the quotient.

    _____

    B. Solve the number sentence you wrote for Part A. Explain your answer.

    _____

    _____

    _____

Common Core State Standards:
4.NBT.1, 4.NBT.2

Read and Write Whole Numbers

Getting the Idea

A **whole number** can be written in different forms:

base-ten numeral: 134,582

number name: one hundred thirty-four thousand, five hundred eighty-two

expanded form: 100,000 + 30,000 + 4,000 + 500 + 80 + 2

Place value is the value of a digit in a number based on its location.

You can use a place-value chart to find the value of each digit.

The digit 3 is in the ten thousands place. It has a value of 30,000.

Hundred Thousands	Ten Thousands	Thousands	,	Hundreds	Tens	Ones
1	3	4	,	5	8	2

The value of a digit is 10 times the value of the digit to its right.

The models below represent the number 2,222.

The 2 in the thousands place has a value of 2,000.
That is 10 times the value of the 2 in the hundreds place.

The 2 in the hundreds place has a value of 200.
That is 10 times the value of the 2 in the tens place.

The 2 in the tens place has a value of 20.
That is 10 times the value of the 2 in the ones place.

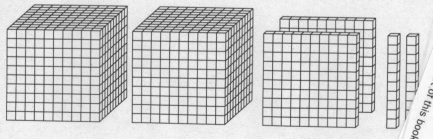

2,000 ÷ 200 = 10

200 ÷ 20 = 10

20 ÷ 2 = 10

Example 1

A singer on a show received 45,698 votes from viewers. What is the value of the 5 in 45,698?

Strategy   **Use a place-value chart.**

Step 1   Write each digit of the number in a chart.

Ten Thousands	Thousands	,	Hundreds	Tens	Ones
4	5	,	6	9	8

Step 2   Find the value of the 5.

The 5 is in the thousands place.

The value of the 5 is 5,000.

Solution   **The value of the 5 in 45,698 is 5,000.**

Example 2

A company has 312,775 employees. What is the number name for 312,775?

Strategy   **Use place value. Look at the comma.**

Step 1   Find the value of the digits before the comma.

There are 312 thousands.

Write *three hundred twelve thousand.*

Step 2   Find the value of the digits after the comma.

There are 775.

Write *seven hundred seventy-five.*

Step 3   Write the number name.

Put a comma after the thousands.

three hundred twelve thousand, seven hundred seventy-five

Solution   **The number name for 312,775 is *three hundred twelve thousand, seven hundred seventy-five.***

Example 3

How can you write the number 954,362 in expanded form?

Strategy **Use a place-value chart.**

Step 1 Write each digit of the number in a chart.

Hundred Thousands	Ten Thousands	Thousands	,	Hundreds	Tens	Ones
9	5	4	,	3	6	2

Step 2 Write the value of each digit.

9 hundred thousands = 900,000

5 ten thousands = 50,000

4 thousands = 4,000

3 hundreds = 300

6 tens = 60

2 ones = 2

Step 3 List the values. Use a + between each value.

900,000 + 50,000 + 4,000 + 300 + 60 + 2

Solution **In expanded form, 54,362 is 900,000 + 50,000 + 4,000 + 300 + 60 + 2.**

Coached Example

Laura bought a new home for $239,807.

Write the expanded form and the number name for this dollar amount.

Write the value of each digit.

What is the value of the 2? _200,000_

What is the value of the 3? _30,000_

What is the value of the 9? _9,000_

What is the value of the 8? _800_

What is the value of the 0? _10_

What is the value of the 7? _7_

The expanded form of 239,807 is _____.

Find the value of the digits before the comma.

There are _____ thousands.

Write the value in words.

_____

Find the value of the digits after the comma.

There are _____.

Write the value in words. _____

Place a comma after the thousands.

The number name for 239,807 is

_____.

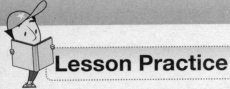

Lesson Practice

Choose the correct answer.

1. A toy store had 27,436 customers last week. Which digit is in the thousands place in 27,436?

   A. 2

   B. 3

   C. 4

   D. 7

2. Which is another way to show this number?

   thirteen thousand,
   one hundred nineteen

   A. 13,119

   B. 13,190

   C. 13,191

   D. 13,911

3. In 2009, there were 15,095 airports in the United States. What is the value of the digit 9 in 15,095?

   A. 9,000

   B. 900

   C. 90

   D. 9

4. Which number has the digit 6 in the thousands place and in the tens place?

   A. 13,662

   B. 44,668

   C. 76,361

   D. 86,629

5. Which is the number name for 240,048?

   A. two hundred four thousand, four hundred eight

   B. two hundred four thousand, forty-eight

   C. two hundred forty thousand, four hundred eight

   D. two hundred forty thousand, forty-eight

6. How many times greater is a digit in the thousands place than that same digit in the hundreds place?

   A. 1

   B. 10

   C. 100

   D. 1,000

7. A phone company serves 475,931 households. Which shows the expanded form of 475,931?

   A. 400,000 + 70,000 + 50,000 + 9,000 + 300 + 10

   B. 400,000 + 70,000 + 5,000 + 900 + 30 + 1

   C. 40,000 + 7,000 + 5,000 + 900 + 30 + 1

   D. 100,000 + 30,000 + 9,000 + 500 + 70 + 4

8. Which shows the expanded form of 945,025?

   A. 900,000 + 40,000 + 5,000 + 20 + 5

   B. 900,000 + 40,000 + 5,000 + 200 + 50

   C. 90,000 + 40,000 + 50,000 + 20 + 5

   D. 900,000 + 40,000 + 500 + 20 + 5

9. A school collected 15,838 plastic bottles to be recycled.

   A. Write the number of bottles in expanded form.

   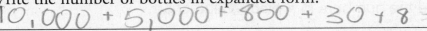
   10,000 + 5,000 + 800 + 30 + 8 =

   B. How many times greater is the 8 in the hundreds place than the 8 in the ones place? Explain your answer.

Common Core State Standard:
4.NBT.2

Compare and Order Whole Numbers

Getting the Idea

You can compare numbers using place value.

Use these symbols to compare numbers.

The symbol > means **is greater than**.

The symbol < means **is less than**.

The symbol = means **is equal to**.

Example 1

Which symbol makes this sentence true? Write >, <, or =.

65,912 ⟩ 65,879

Strategy **Use a place-value chart. Start with the digits in the greatest place.**

| Step 1 | Write the numbers in a place-value chart. |

Ten Thousands	Thousands	,	Hundreds	Tens	Ones
6	5	,	9	1	2
6	5	,	8	7	9

Step 2 Compare the digits in the ten thousands place.

Both numbers have 6 in the ten thousands place.

Compare the next greatest place.

Step 3 Compare the digits in the thousands place.

Both numbers have 5 in the thousands place.

Compare the next greatest place.

Step 4 Compare the digits in the hundreds place.

9 hundreds are greater than 8 hundreds.

So, 65,912 is greater than 65,879.

Step 5 Choose the correct symbol.

> means is greater than.

Solution **65,912 > 65,879**

Example 2

A city's budget for maintaining its parks for one year was $718,325. The town spent $718,352 that year. Did the city spend more or less than the budgeted amount?

$718,325 ⬉ $718,352

Strategy	**Line up the numbers on the ones place.**
	Then compare the digits from left to right.

718,325

718,352

Step 1 Compare the digits in the hundred thousands place.

Since 7 = 7, compare the next greatest place.

Step 2 Compare the digits in the ten thousands place.

Since 1 = 1, compare the next greatest place.

Step 3 Compare the digits in the thousands place.

Since 8 = 8, compare the next greatest place.

Step 4 Compare the digits in the hundreds place.

Since 3 = 3, compare the next greatest place.

Step 5 Compare the digits in the tens place.

Since 2 < 5, then 718,325 < 718,352.

Solution **The city spent more than the budgeted amount.**

When you order numbers, find the greatest number and the least number. Compare two numbers at a time, and then order the numbers.

Example 3

Order the following numbers from least to greatest.

   527,877   528,371   527,918

Strategy   **Line up the numbers on the ones place.**
               Start comparing the digits in the greatest place.

Step 1   Compare the digits in the hundred thousands place.
               All the digits are 5s.

Step 2   Compare the digits in the ten thousands place.
               All the digits are 2s.

Step 3   Compare the digits in the thousands place.
               Since $8 > 7$, then $528,371 > 527,877$ and $527,918$.
               528,371 is the greatest number.

Step 4   For the remaining numbers, compare the digits in the hundreds place.
               Since $8 < 9$, then $527,877 < 527,918$.

Solution   **The order of the numbers from least to greatest is 527,877;**
               527,918; 528,371.

Coached Example

Which symbol makes this sentence true? Write >, <, or =.

693,041 ◯ 693,582

Use a place-value chart. Write the numbers in the chart.

Hundred Thousands	Ten Thousands	Thousands	,	Hundreds	Tens	Ones
6	9	3		0	4	1
6	9	3		5	8	2

Compare the digits in the hundred thousands place.

Are the digits in the hundred thousands place the same? _Yes_

Compare the digits in the ten thousands place.

Are the digits in the ten thousands place the same? _Yes_

Compare the digits in the thousands place.

Are the digits in the thousands place the same? _Yes_

Compare the digits in the hundreds place.

Are the digits in the hundreds place the same? _No_

0 hundreds is _____less_____ than 5 hundreds.

So, 693,041 is _____less_____ than 693,582.

Which symbol should you use? _geater than_

693,041 ◯ 693,582

Lesson Practice

Choose the correct answer.

1. Which sentence is true?

   A. 78,412 > 79,421

   B. 67,905 < 76,905

   C. 19,058 > 21,037

   D. 52,915 < 52,836

2. Which symbol makes this sentence true?

   237,352 ◯ 237,452

   A. >

   B. <

   C. =

   D. +

3. Which list orders the numbers from greatest to least?

   A. 67,358   72,185   72,581

   B. 72,581   67,358   72,185

   C. 72,581   72,185   67,358

   D. 72,185   72,581   67,358

4. The table shows the seating capacities of the stadiums of four baseball teams.

   Stadium Seating Capacity

Team	Seating Capacity
Diamondbacks	49,033
Orioles	48,876
Rangers	49,200
Twins	48,678

   Which team's stadium has the greatest seating capacity?

   A. Diamondbacks

   B. Orioles

   C. Rangers

   D. Twins

5. The numbers are ordered from greatest to least. One number is missing.

   582,364   __?__   578,264

   Which number is missing?

   A. 573,095

   B. 575,195

   C. 578,263

   D. 578,493

6. Which number is greater than 128,278 and less than 129,384?

   A. 128,209

   B. 128,728

   C. 129,394

   D. 129,438

7. Which digit makes this sentence true?

$$48,185 < 4\boxed{},242$$

   A. 8

   B. 7

   C. 6

   D. 0

8. The table shows the number of ice cream cones sold at Bennie's Ice Cream Parlor each year for four years.

Ice Cream Cone Sales

Year	Ice Cream Cones
2007	11,296
2008	11,474
2009	12,107
2010	12,044

   A. In which year was the greatest number of ice cream cones sold? Explain how you found your answer.

The year 2009 was the greatest year They sold ice cearm bars. I knew because in the hundreds place

   B. In which year was the least number of ice cream cones sold? Explain how you found your answer.

Multiplication Facts

Common Core State Standards:
4.OA.1, 4.OA.2, 4.OA.3

Getting the Idea

You can **multiply** to find the total number of equal groups.

Here are the parts in a multiplication sentence.

$$5 \times 4 = 20$$

factor   factor   product

You can use an array to show multiplication. An **array** has the same number of objects in each row.

Example 1

What multiplication sentence does this array show?

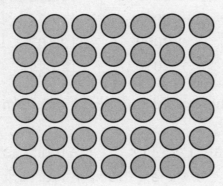

Strategy   **Count the number of rows. Then count the number of counters in each row.**

Step 1   Count the number of rows and the number of counters in each row.

There are 6 rows and 7 counters in each row.

Step 2   Find the number of counters in all.

There are 42 counters in all.

Step 3   Write the number sentence.

The factors are 6 and 7 and the product is 42.

$6 \times 7 = 42$

Solution   **The array shows $6 \times 7 = 42$.**

You can also use repeated addition to solve a multiplication problem. Repeated addition is adding the same number over and over again. Multiplication is a shortcut for repeated addition.

To find 3 × 4, you can add 4 three times: 4 + 4 + 4 = 12

Repeated addition is similar to skip counting.

To find 3 × 4, you can skip count by 4 three times: 4, 8, 12

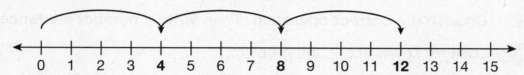

Example 2

Stephanie baked 4 pies for her school's bake sale. She cut each pie into 8 slices. How many slices did Stephanie make in all?

Strategy   **Use repeated addition.**

Step 1   Write the multiplication sentence for the problem.

She made 4 pies. Each pie has 8 slices.

Find 4 groups of 8.

4 × 8 = ☐

Step 2   Write the repeated addition for the multiplication sentence.

4 × 8 is the same as adding 8 four times.

4 × 8 = 8 + 8 + 8 + 8

Step 3   Find the sum.

8 + 8 + 8 + 8 = 32

Solution   **Stephanie made 32 slices of pie in all.**

A **variable** is a letter or symbol used to represent a value that is unknown.

You can use a variable to represent an unknown value in a number sentence.

Example 3

A shirt costs $5. A jacket costs 6 times as much as the shirt.
How much does the jacket cost?

Strategy   **Choose the correct operation. Then write a number sentence.**

Step 1   Look for key words in the problem.

The key words "times as much" mean you should multiply.

Step 2   What number do you multiply?

Think: a jacket costs 6 times as much as a $5 shirt.

You need to multiply 6×5.

Step 3   Write the number sentence. Use n for the cost of the jacket.

$6 \times 5 = n$

Step 4   Solve the number sentence.

Think: $6 \times 5 = 5 + 5 + 5 + 5 + 5 + 5$

$5 + 5 + 5 + 5 + 5 + 5 = 30$

Solution   **The jacket costs $30.**

You can use a multiplication table to help you learn basic multiplication facts.
The factors are along the first column and the top row.
The box where the row and the column meet is the product.

Columns

×	0	1	2	3	4	5	6	7	8	9	10	11	12
0	0	0	0	0	0	0	0	0	0	0	0	0	0
1	0	1	2	3	4	5	6	7	8	9	10	11	12
2	0	2	4	6	8	10	12	14	16	18	20	22	24
3	0	3	6	9	12	15	18	21	24	27	30	33	36
4	0	4	8	12	16	20	24	28	32	36	40	44	48
5	0	5	10	15	20	25	30	35	40	45	50	55	60
6	0	6	12	18	24	30	36	42	48	54	60	66	72
7	0	7	14	21	28	35	42	49	56	63	70	77	84
8	0	8	16	24	32	40	48	56	64	72	80	88	96
9	0	9	18	27	36	45	54	63	72	81	90	99	108
10	0	10	20	30	40	50	60	70	80	90	100	110	120
11	0	11	22	33	44	55	66	77	88	99	110	121	132
12	0	12	24	36	48	60	72	84	96	108	120	132	144

Rows

Example 4

Naomi is 11 years old. Naomi's great grandmother is 9 times Naomi's age.
How old is Naomi's great grandmother?

Strategy   **Use a multiplication table.**

Step 1   Write the multiplication sentence for the problem.

Naomi is 11. Her great grandmother is 9 times Naomi's age.

Find 11 groups of 9.

$11 \times 9 = n$

Step 2   Look at the 11s column.

Step 3   Find the 9s row.

Step 4   Find the box where the 11s column and the 9s row meet.

The number 99 is in the box.

So, $11 \times 9 = 99$.

Solution   **Naomi's great grandmother is 99 years old.**

Coached Example

Kate gives her dog 3 biscuits each day.
How many biscuits does Kate give her dog in 7 days?

Write the multiplication sentence for the problem.

Find ___3___ groups of ___7___ biscuits.

___3___ × ___7___ = [21]

Use the multiplication table.

Find the ___7___s column.

Find the ___3___s row.

Find the box where the row and the column meet.

The number ___21___ is in the box.

So, ___3___ × ___7___ = ___21___.

Kate gives her dog ___21___ biscuits in 7 days.

Lesson Practice

Choose the correct answer.

1. Multiply.

    $7 \times 5 = \square$

    A. 35

    B. 36

    C. 42

    D. 49

2. Which multiplication sentence does this repeated addition show?

    $3 + 3 + 3 + 3 + 3 + 3$

    A. $3 \times 3 = \square$

    B. $5 \times 3 = \square$

    C. $6 \times 3 = \boxed{18}$

    D. $7 \times 3 = \square$

3. Multiply.

    $12 \times 6 = \square$

    A. 6

    B. 18

    C. 60

    D. 72

    $\begin{array}{r} 12 \\ \times\ \ 6 \\ \hline 72 \end{array}$

4. A small potted plant costs $4. A large potted plant costs 5 times as much. How much does the large potted plant cost?

    A. $9

    B. $16

    C. $20

    D. $25

5. A blue shirt costs $7. A green shirt costs 3 times as much. How much does the green shirt cost?

    A. $10

    B. $21

    C. $24

    D. $28

6. For a park cleanup, there were 6 volunteers on each team. There were 4 teams. How many volunteers helped clean up the park?

    A. 36

    B. 32

    C. 28

    D. 24

7. A dime is worth 10 cents. How much are 7 dimes worth?

    A.   7 cents

    B. 17 cents

    C. 35 cents

    D. 70 cents

8. Which multiplication fact has the greatest product?

    A. $6 \times 9 = 54$

    B. $5 \times 10 = 50$

    C. $4 \times 11 = 44$

    D. $3 \times 12 = 36$

9. A school bus has 12 rows of seats. Each row can fit 4 students.

    A. Write a number sentence that shows how many students can sit on the bus.

    _____

    B. How many students can sit on the bus? Explain how you found your answer.

    _____

    _____

    _____

Multiply Greater Numbers

Common Core State Standards:
4.NBT.5, 4.OA.3

Getting the Idea

You can multiply greater numbers by using basic facts and regrouping.
Sometimes using models can help you multiply.

Example 1

Multiply.

$3 \times 24 = \boxed{}$

Strategy   **Use an area model.**

Step 1   Use an area model of 3 rows of 24 squares.

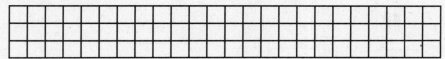

Step 2   Count the number of rows. Count the number of squares in each row.

There are 3 rows.

Each row has 24 squares.

Step 3   Find the total number of squares.

3 groups of 24 equals 72.

Solution   **$3 \times 24 = 72$**

Example 2

Washington Elementary School has 4 sections of seats in the cafeteria.
Each section has 48 seats. How many seats in all are in the cafeteria?

Strategy   **Multiply by the ones. Then multiply by the tens.**

Step 1   Write a multiplication sentence for the problem.

The cafeteria has 4 sections of seats.

There are 48 seats in each section.

There are *n* total sections.

$4 \times 48 = n$

Step 2 Write the problem in vertical form.

$$\begin{array}{r} 48 \\ \times\ 4 \\ \hline \end{array}$$

Step 3 Multiply the ones: $4 \times 8 = 32$.

Write the 2 and regroup the 3 tens.

$$\begin{array}{r} 3 \\ 48 \\ \times\ 4 \\ \hline 2 \end{array}$$

Step 4 Multiply the tens: $4 \times 4 = 16$.

Add the regrouped tens: $16 + 3 = 19$.

Write the 19.

$$\begin{array}{r} 3 \\ 48 \\ \times\ 4 \\ \hline 192 \end{array}$$

Solution **There are 192 seats in all in the cafeteria.**

Example 3

Tasha bought 7 rolls of paper streamers for the school dance. Each roll is 328 inches long. How many inches of paper streamers did Tasha buy in all?

Strategy **Multiply each place by 7, regrouping when necessary.**

Step 1 Write the problem vertically.

$$\begin{array}{r} 328 \\ \times\ \ 7 \\ \hline \end{array}$$

Step 2 Multiply the ones.

$7 \times 8 = 56$

Write the 6 and regroup 5 tens.

$$\begin{array}{r} 5 \\ 328 \\ \times\ \ 7 \\ \hline 6 \end{array}$$

Step 3   Multiply the tens and add the regrouped tens.

$7 \times 2 = 14$

$14 + 5 = 19$

Write the 9 and regroup 1 hundred.

```
  15
 328
×   7
──────
  96
```

Step 4   Multiply the hundreds and add the regrouped hundreds.

$7 \times 3 = 21$

$21 + 1 = 22$

Write the 22 hundreds.

```
   15
  328
×    7
──────
2,296
```

Solution   **Tasha bought 2,296 inches of paper streamers.**

Example 4

Mrs. Rivera earned $1,635 each week for 4 weeks. How much money did she earn in all?

Strategy   **Multiply each place by 4, regrouping when necessary.**

Step 1   Write the multiplication sentence for the problem.

She earned $1,635 each week for 4 weeks.
Use n for the unknown product.

$1,635 \times 4 = n$

Step 2   Write the problem vertically.

```
 1,635
×    4
```

Step 3   Multiply the ones.

$$4 \times 5 = 20$$

Write the 0 and regroup 2 tens.

$$
\begin{array}{r}
2 \\
1,635 \\
\times \quad 4 \\
\hline
0
\end{array}
$$

Step 4   Multiply the tens and add the regrouped tens.

$$4 \times 3 = 12$$

$$12 + 2 = 14$$

Write the 4 and regroup 1 hundred.

$$
\begin{array}{r}
1\,2 \\
1,635 \\
\times \quad 4 \\
\hline
40
\end{array}
$$

Step 5   Multiply the hundreds and add the regrouped hundreds.

$$4 \times 6 = 24$$

$$24 + 1 = 25$$

Write the 5 hundreds and regroup 2 thousands.

$$
\begin{array}{r}
2\,1\,2 \\
1,635 \\
\times \quad 4 \\
\hline
540
\end{array}
$$

Step 6   Multiply the thousands and add the regrouped thousands.

$$4 \times 1 = 4$$

$$4 + 2 = 6$$

Write the 6 thousands.

$$
\begin{array}{r}
2\,1\,2 \\
1,635 \\
\times \quad 4 \\
\hline
6,540
\end{array}
$$

Solution   **Mrs. Rivera earned $6,540 in all.**

When multiplying two-digit numbers, first multiply a factor by the ones digit of the other factor. Then multiply the same factor by the tens digit of the other factor. Finally, add the partial products to find the final product.

Example 5

Multiply.

$$45 \times 28 = \boxed{}$$

Strategy   **Multiply each place value. Regroup when necessary.**

Step 1   Write the problem in vertical form.

$$\begin{array}{r} 45 \\ \times\, 28 \\ \hline \end{array}$$

Step 2   Multiply 45 by the ones digit of 28:

8 ones × 45. Regroup.

$$\begin{array}{r} 4 \\ 45 \\ \times\, 28 \\ \hline 360 \end{array}$$ ← partial product

Step 3   Multiply 45 by the tens digit of 28: 2 tens × 45.

Write a 0 in the ones place because you are multiplying the tens.

Regroup.

$$\begin{array}{r} 1 \\ 45 \\ \times\, 28 \\ \hline 360 \\ 900 \end{array}$$ ← partial product

Step 4   Add the partial products.

$$\begin{array}{r} 45 \\ \times\, 28 \\ \hline 360 \\ +\, 900 \\ \hline 1260 \end{array}$$

Solution $45 \times 28 = 1{,}260$

Coached Example

A student ticket to a theme park costs $34. A class of 26 fourth-grade students went to the theme park. How much did the tickets for the students cost in all?

Write the multiplication sentence for the problem.

A student ticket costs $ ___34___.

A class of ___26___ students went to the park.

Find ___34___ × ___26___ = ☐.

Write the problem in vertical form.

$$
\begin{array}{r}
\overset{2}{34} \\
\times 26 \\
\hline
204 \\
680 \\
\hline
884
\end{array}
$$

Multiply 34 by the ___1___ digit of 26.

___6___ ones × 34. Regroup.

What is the first partial product? ___20 4___

Multiply 34 by the ___1___ digit of 26.

___20___ tens × 34. Regroup.

What is the second partial product? ___680___

Add the two partial products.

___680___ + ___204___ = ___884___

The tickets for the students cost $___884___ in all.

Lesson Practice

Choose the correct answer.

1. Multiply.

    48
    × 8

    A. 384
    B. 364
    C. 352
    D. 324

2. Multiply.

    $9 \times 37 =$ ▢

    A. 323
    B. 333
    C. 343
    D. 433

3. A tour group of 68 people has an overnight stay at a hotel. Each person will receive a 3-pancake breakfast. How many pancakes will the hotel serve to the tour group?

    A. 184
    B. 194
    C. 204
    D. 214

4. Hiro brushes his teeth 3 times each day. How many times will he brush his teeth in a month with 31 days?

    A. 11
    B. 33
    C. 93
    D. 930

5. Multiply.

    74
    × 11

    A. 85
    B. 747
    C. 814
    D. 7,474

6. Multiply.

    $4,962 \times 6 =$ ▢

    A. 27,401
    B. 27,410
    C. 29,766
    D. 29,772

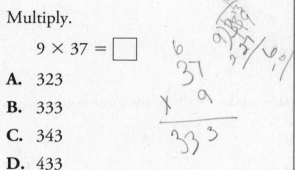
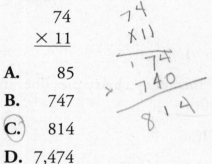
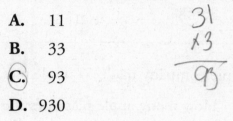

7. An auditorium has 68 rows of seats. There are 56 seats in each row. How many seats are there in all?

   Ⓐ 3,808

   B. 3,768

   C. 3,408

   D. 3,048

$$\begin{array}{r} 356 \\ \times 68 \\ \hline 448 \\ 3360 \\ \hline 3{,}808 \end{array}$$

8. A baseball cap and T-shirt set costs $9. What will be the total cost for 1,086 sets?

   A. $8,688

   B. $9,765

   Ⓒ $9,774

   D. $10,317

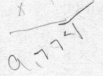

9. A pie company made 57 apple pies and 38 cherry pies each day for 14 days.

   A. How many apple pies does the company make in all? Show your work.

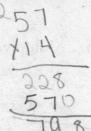

$$\begin{array}{r} 57 \\ \times 14 \\ \hline 228 \\ 570 \\ \hline 798 \end{array}$$

   B. How many cherry pies does the company make in all? Show your work.

$$\begin{array}{r} 38 \\ \times 14 \\ \hline 152 \\ 380 \\ \hline 532 \end{array}$$

Multiplication Properties

Common Core State Standards:
4.NBT.5

Getting the Idea

There are some mathematical properties that can help make multiplication easier for you. Properties are rules.

> **Commutative Property of Multiplication**
> The order of the factors can be changed.
> The product does not change.
>
> $12 \times 18 = 18 \times 12$
>
> $216 = 216$

Example 1

Which number makes this number sentence true?

$13 \times \boxed{} = 23 \times 13$

Strategy   **Use the commutative property of multiplication.**

Step 1   Look at the number sentence.

The left side of the equal sign shows $13 \times \boxed{}$.

The right side of the equal sign shows 23×13.

The equal sign means that they have the same product.

Step 2   Think about the commutative property of multiplication.

The order of the factors does not change the product.

$13 \times 23 = 23 \times 13$

Solution   **The number 23 makes the number sentence true.**

> ## Multiplicative Identity Property of 1
>
> When you multiply any number by 1, the product is that number.
>
> $$1 \times 57 = 57$$

Example 2

Which number makes this number sentence true?

$$\boxed{} \times 1 = 82$$

Strategy   **Use the multiplicative identity property of 1.**

Step 1   Look at the number sentence.

     The left side of the equal sign shows $\boxed{} \times 1$.

     The right side of the equal sign shows 82.

Step 2   Use the multiplicative identity property of 1.

     Any number multiplied by 1 is that number.

     Since one of the factors is 1, the other factor is 82.

     $82 \times 1 = 82$

Solution   **The number 82 makes the number sentence true.**

Associative Property of Multiplication
Factors can be grouped in different ways.
The product will be the same.

$$(12 \times 7) \times 14 = 12 \times (7 \times 14)$$

$$84 \times 14 = 12 \times 98$$

$$1{,}176 = 1{,}176$$

Example 3

Multiply.

$$8 \times (12 \times 10) = \boxed{}$$

Strategy   **Use the associative property of multiplication.**

Step 1   Think about the associative property of multiplication.
              The grouping of the factors does not change the product.

Step 2   Regroup the factors.
              $$8 \times (12 \times 10) = (8 \times 12) \times 10$$

Step 3   Use mental math to multiply.
              Multiply inside the parentheses. Then find the final product.
              $$(8 \times 12) \times 10 = \boxed{}$$
              $$96 \times 10 = 960$$

Solution $8 \times (12 \times 10) = 960$

Coached Example

A bag has 5 packets of jellybeans. Each packet has 14 jellybeans. Joey bought 2 bags. How many jellybeans did Joey buy in all?

5 × 14 × 2 = ☐

Use the _____ property of multiplication to change the order of the factors.

_____ × _____ × _____ = ☐

Use the _____ property of multiplication to group the factors.

(_____ × _____) × _____ = ☐

Multiply inside the parentheses.

(_____) × _____ = ☐

Multiply that factor and the other factor.

_____ × _____ = _____

So, 5 × 14 × 2 = _____

Joey bought _____ jellybeans in all.

Lesson Practice

Choose the correct answer.

1. What is the missing number in this sentence?

   $31 \times 43 = 43 \times \boxed{}$

   A. 1

   B. 12

   C. 31

   D. 43

2. Which correctly shows the commutative property of multiplication?

   A. $85 \times 0 = 0$

   B. $7 \times 13 = 91$

   C. $8 \times 17 = 17 \times 8$

   D. $1 \times 39 = 39$

3. What is the missing number in this sentence?

   $68 \times \boxed{} = 68$

   A. 0

   B. 1

   C. 67

   D. 68

4. Which correctly shows the multiplicative identity property of 1?

   A. $27 \times 3 = 3 \times 27$

   B. $29 \times 1 = 29$

   C. $0 \times 18 = 0$

   D. $2 \times (35 \times 4) = (2 \times 35) \times 4$

5. Which number sentence is true?

   A. $32 \times 4 = 4 \times 32$

   B. $32 \times 4 = 4 + 32$

   C. $32 \times 4 = 324$

   D. $32 \times 4 = 32 \div 4$

6. What is the missing number in this sentence?

   $12 \times (5 \times 13) = (\boxed{} \times 5) \times 13$

   A. 5

   B. 7

   C. 12

   D. 13

7. Multiply.

$$10 \times (9 \times 21) = \boxed{}$$

A.   90

B.   189

C.   210

D.   1,890

8. Multiply.

$$(18 \times 5) \times 12 = \boxed{}$$

A.   60

B.   90

C.   216

D.   1,080

9. A bus has 12 rows of seats. Each row can fit 6 passengers.

Mr. Kane ordered 5 buses to take people to a baseball game.

$$12 \times 6 \times 5 = \boxed{}$$

A. Use the commutative property of multiplication to change the order of the factors.

_____

B. Use the associative property of multiplication to group the factors.

_____

C. How many passengers in all can Mr. Kane take to the baseball game?

_____

Common Core State Standards:
4.OA.3, 4.NBT.5

Distributive Property of Multiplication

Getting the Idea

The distributive property of multiplication can help you multiply numbers using mental math. The property uses the expanded form of numbers.

Area models can help you understand the distributive property of multiplication.

Distributive Property of Multiplication

When you multiply a number by a sum, you can multiply the number by each addend of the sum and then add the products.

$$5 \times 14$$

$$5 \times (10 + 4)$$

$$(5 \times 10) + (5 \times 4)$$

$$50 \quad + \quad 20 = 70$$

$$5 \times 14 = 70$$

$$5 \times 10 = 50 \qquad 5 \times 4 = 20$$

Example 1

Fred has 3 shelves of books. Each shelf has 18 books. How many books in all are on the shelves?

Strategy   **Use the distributive property of multiplication and mental math.**

Step 1   Write the multiplication sentence for the problem.

There are 3 shelves. There are 18 books on each shelf.
There are x books in all.

$$3 \times 18 = x$$

Step 2   Express 18 in expanded form.

$$18 = 10 + 8$$

Step 3   Rewrite the sentence with 18 in expanded form.

$$3 \times 18 = 3 \times (10 + 8)$$

Step 4   Distribute the 3 to each addend.

$$3 \times (10 + 8) = (3 \times 10) + (3 \times 8)$$

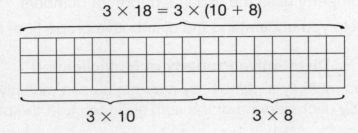

$$3 \times 18 = 3 \times (10 + 8)$$

$$3 \times 10 \qquad 3 \times 8$$

Step 5   Find each product.

$$(3 \times 10) + (3 \times 8) = x$$
$$30 \quad + \quad 24 \quad = x$$

Step 6   Add the products.

$$30 + 24 = 54$$

Solution   **There are 54 books in all on the shelves.**

Example 2

Multiply.

$$12 \times 34 = \boxed{}$$

Strategy   **Use the distributive property and mental math.**

Step 1   Express 34 in expanded form.

$$34 = 30 + 4$$

Step 2   Rewrite the sentence with 34 in expanded form.

$$12 \times 34 = 12 \times (30 + 4)$$

Step 3   Distribute the 12 to each addend.

$$12 \times (30 + 4) = (12 \times 30) + (12 \times 4)$$

Step 4	Find each product.

$$(12 \times 30) + (12 \times 4) = \boxed{}$$
$$360 \quad + \quad 48 \quad = \boxed{}$$

Step 5	Add the products.

$$360 + 48 = 408$$

Solution $12 \times 34 = 408$

Example 3

A Blu-Ray DVD costs $25. Ms. Ely ordered 15 Blu-Ray DVDs. How much did Ms. Ely spend in all on the DVDs?

Strategy   **Use the distributive property and mental math.**

Step 1	Write the multiplication sentence for the problem.

> She bought 15 DVDs. Each DVD cost $25.
> She spent n dollars in all.
>
> $15 \times \$25 = n$

Step 2	Express 25 in expanded form.

$$25 = 20 + 5$$

Step 3	Rewrite the sentence with 25 in expanded form.

$$15 \times 25 = 15 \times (20 + 5)$$

Step 4	Distribute the 15 to each addend.

$$15 \times (20 + 5) = (15 \times 20) + (15 \times 5)$$

Step 5	Find each product.

$$(15 \times 20) + (15 \times 5) = n$$
$$300 \quad + \quad 75 \quad = n$$

Step 6	Add the products.

$$300 + 75 = 375$$

Solution   **Ms. Ely spent $375 in all.**

Coached Example

Monroe Elementary School has 32 classrooms. Each classroom has 24 students. How many students in all are at the school?

Write the multiplication sentence for the problem.

There are _____ classrooms.

There are _____ students in each class.

There are n students in all.

_____ $\times$ _____ = _____

Use the distributive property of multiplication.

Express 24 in expanded form.

24 = _____ + _____

Rewrite the sentence with 24 in expanded form.

$32 \times 24 = 32 \times ($ _____ + _____ $)$

Distribute 32 to each addend.

$32 \times ($ _____ + _____ $) = (32 \times$ _____ $) + (32 \times$ _____ $)$

Find each product.

$(32 \times$ _____ $) + (32 \times$ _____ $) = n$

_____ + _____ = n

Add the products.

_____ + _____ = _____

There are _____ students in all at the school.

Lesson Practice

Choose the correct answer.

1. Which is true?

   A. $3 \times 78 = (3 \times 70) \times (3 \times 8)$

   B. $3 \times 78 = (3 \times 70) + (3 \times 8)$

   C. $3 \times 78 = (3 + 70) \times (3 + 8)$

   D. $3 \times 78 = (3 + 70) + (3 + 8)$

2. Which is true?

   A. $64 \times 14 = (64 \times 10) \times (64 \times 4)$

   B. $64 \times 14 = (64 \times 10) + (64 \times 4)$

   C. $64 \times 14 = (64 + 10) \times (64 + 4)$

   D. $64 \times 14 = (64 + 10) + (64 + 4)$

3. Which is true?

   A. $52 \times 23 = (52 \times 20) + (52 \times 3)$

   B. $52 \times 23 = (50 \times 20) + (2 \times 3)$

   C. $52 \times 23 = (52 \times 20) + (2 \times 3)$

   D. $52 \times 23 = (52 \times 20) \times (52 \times 3)$

4. Multiply.

   $16 \times 24 = \boxed{}$

   A. 96

   B. 324

   C. 326

   D. 384

5. Multiply.

   $23 \times 23 = \boxed{}$

   A. 115

   B. 246

   C. 529

   D. 1,024

6. A club charges $26 for a one-year membership. The club has 62 members. How much does the club collect in membership fees each year?

   A. $208

   B. $1,560

   C. $1,612

   D. $2,408

7. Rosa bought 15 cases of water for a school fair. Each case has 24 bottles. How many bottles of water did Rosa buy?

    A. 200

    B. 260

    C. 300

    D. 360

8. A manatee's heart normally beats about 55 times a minute. How many times does a manatee's heart beat in 60 minutes?

    A. 33,000

    B. 3,300

    C. 3,000

    D. 330

9. The art teacher bought 32 boxes of crayons. Each box has 64 crayons.

    A. Write a number sentence to find how many crayons the art teacher bought in all.

    _____

    B. Use the distributive property of multiplication to find the total number of crayons. Show your work.

    _____

    _____

    _____

Division Facts

Common Core State Standards:
4.0A.2, 4.0A.3

Getting the Idea

You can **divide** to find the number of equal groups or the number in each group. Here are the parts in a division sentence.

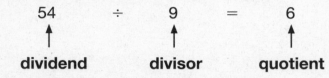

You can use an array to show division.

Example 1

Michael bagged 32 cans of soup in 4 bags. Each bag has the same number of cans. How many cans are in each bag?

Strategy   **Make an array.**

Step 1   Write the division sentence for the problem.

There are 32 cans in all. There are 4 bags.
There are n cans in each bag.

$32 \div 4 = n$

Step 2   Use 32 counters. Put the counters in 4 equal rows.

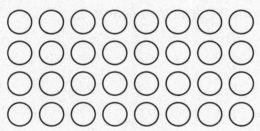

Step 3   Find the number of counters in each row.

There are 8 counters in each row.

Solution   **There are 8 cans in each bag.**

You can use repeated subtraction to solve a division problem. Start with the dividend. Subtract the divisor over and over until you reach 0. The number of times you subtracted is the quotient.

For example, to find $18 \div 6$:

Start at 18. Subtract 6 until you reach 0.

$18 - 6 = 12$

$12 - 6 = 6$

$6 - 6 = 0$

6 was subtracted 3 times. So, $18 \div 6 = 3$.

Example 2

Sandra bakes 24 cookies that she shares equally with 3 of her friends. How many cookies will each friend get?

Strategy	**Use repeated subtraction.**
Step 1	Write the division sentence for the problem.
	There are 24 cookies in all.
	There are 4 friends.
	Each friend will get n cookies.
	$24 \div 4 = n$
Step 2	Subtract 4 from 24 until you reach 0.
	$24 - 4 = 20$
	$20 - 4 = 16$
	$16 - 4 = 12$
	$12 - 4 = 8$
	$8 - 4 = 4$
	$4 - 4 = 0$
Step 3	Count the number of times 4 was subtracted.
	4 was subtracted 6 times.
	$24 \div 4 = 6$
Solution	**Each friend gets 6 cookies.**

Division is the opposite, or the **inverse operation**, of multiplication.
You can use a multiplication table to solve division.

×	0	1	2	3	4	5	6	7	8	9	10	11	12
0	0	0	0	0	0	0	0	0	0	0	0	0	0
1	0	1	2	3	4	5	6	7	8	9	10	11	12
2	0	2	4	6	8	10	12	14	16	18	20	22	24
3	0	3	6	9	12	15	18	21	24	27	30	33	36
4	0	4	8	12	16	20	24	28	32	36	40	44	48
5	0	5	10	15	20	25	30	35	40	45	50	55	60
6	0	6	12	18	24	30	36	42	48	54	60	66	72
7	0	7	14	21	28	35	42	49	56	63	70	77	84
8	0	8	16	24	32	40	48	56	64	72	80	88	96
9	0	9	18	27	36	45	54	63	**72**	81	90	99	108
10	0	10	20	30	40	50	60	70	80	90	100	110	120
11	0	11	22	33	44	55	66	77	88	99	110	121	132
12	0	12	24	36	48	60	72	84	96	108	120	132	144

Example 3

Frank is 72 years old. That is 8 times his granddaughter's age.
What is Frank's granddaughter's age?

Strategy **Use the multiplication table.**

Step 1 Write the division sentence for the problem.

Frank is 72 years old.

72 is 8 times his granddaughter's age.

His granddaughter is n years old.

$72 \div 8 = n$

Step 2 Look at the 8s column.

Step 3 Move down the column and look for 72.

Step 4 Move to the left and look for the row number.

The row is 9.

Solution **Frank's granddaughter is 9 years old.**

Related multiplication and division facts create a **fact family**.

Related facts use the same numbers.

$$5 \times 6 = 30 \qquad 6 \times 5 = 30$$
$$30 \div 5 = 6 \qquad 30 \div 6 = 5$$

Example 4

A log is 49 inches long. Mr. Childs cuts the log into 7 equal pieces.
What is the length of one piece?

Strategy   **Use a related multiplication fact.**

Step 1   Write the division sentence for the problem.

The log is 49 inches long.

It was cut into 7 equal pieces.

Each piece is p inches long.

$$49 \div 7 = p$$

Step 2   Use a related multiplication fact.

Related facts use the same numbers.

Since $7 \times 7 = 49$, then $49 \div 7 = 7$.

Solution   **The length of one piece is 7 inches.**

Coached Example

Ms. Lopez has a total of 35 desks in her classroom. That is 5 times the number of desks in each equal row of desks. How many desks are in each row?

Write a division sentence for the problem.

Use *d* for the number of desks in each row.

There are _____ desks in _____ equal rows.

_____ ÷ _____ = _____

Use a related multiplication fact.

_____ × 5 = 35

Since _____ × 5 = 35, then 35 ÷ 5 = _____.

There are _____ desks in each row.

Lesson Practice

Choose the correct answer.

1. Divide.

   $$45 \div 9 = \boxed{}$$

   A. 5

   B. 6

   C. 7

   D. 8

2. Vanessa used repeated subtraction to solve this division fact.

   $$96 \div 8 = \boxed{}$$

   How many times did she subtract 8 from 96?

   A. 9

   B. 10

   C. 11

   D. 12

3. Which division fact has a quotient of 4?

   A. $20 \div 2 = \boxed{}$

   B. $24 \div 6 = \boxed{}$

   C. $32 \div 4 = \boxed{}$

   D. $36 \div 3 = \boxed{}$

4. Which division fact is related to this multiplication fact?

   $$6 \times 8 = 48$$

   A. $48 \div 2 = 12$

   B. $42 \div 6 = 7$

   C. $48 \div 8 = 6$

   D. $56 \div 8 = 7$

5. Which fact does **not** belong in the same fact family as the others?

   A. $8 \times 2 = 16$

   B. $16 \div 4 = 4$

   C. $2 \times 8 = 16$

   D. $16 \div 2 = 8$

6. Michelle has 54 inches of ribbon. She cut the ribbon into 6-inch pieces. How many pieces of ribbon did Michelle cut?

   A. 6

   B. 7

   C. 8

   D. 9

7. Monica stores 36 paintbrushes in 3 cases. Each case has the same number of paintbrushes. How many paintbrushes are in each case?

A. 9

B. 11

C. 12

D. 16

8. A wooden board is 45 inches long before it is cut into short equal pieces. That is 5 times as long as each short piece. How long is each short piece?

A. 5 inches

B. 6 inches

C. 8 inches

D. 9 inches

9. Sixteen adults and 40 students attended a charity event. The adults and students sat at 8 tables.

A. Each table had the same number of adults. How many adults sat at each table?

_____

B. Each table had the same number of students. How many students sat at each table?

_____

C. How many people in all sat at each table? Show your work.

_____

Divide Greater Numbers

Common Core State Standards:
4.NBT.6, 4.OA.3

Getting the Idea

Division problems can be written in another way.

$$\text{divisor} \overline{)\text{dividend}}^{\text{quotient}}$$

You can use models to help you divide.

Example 1

Divide.

$56 \div 4 = \boxed{}$

Strategy   **Use counters to make an array.**

Step 1   Use 56 counters. Put them in 4 equal rows.

Step 2   Count the number of counters in each row.

There are 14 counters in each row.

Solution   **$56 \div 4 = 14$**

Remember, **inverse operations** are operations that "undo" each other. Addition and subtraction are inverse operations. Multiplication and division are inverse operations. So, you can use multiplication to check a division problem.

For Example 1, you can use $4 \times 14 = 56$ to check $56 \div 4 = 14$.

The product 56 matches the dividend 56, so the answer is correct.

When you divide greater numbers, divide each place of the dividend from left to right.

Example 2

Alex has 72 baseball cards that he wants to store in three cases. If he stores the same number of cards in each case, how many cards will be in each case?

Strategy   **Write a division sentence, then solve.**

Step 1   Write a division sentence for the problem.

He has 72 cards. He has 3 cases. There are *n* cards in each case.

$72 \div 3 = n$

Step 2   Write the problem another way.

$3 \overline{)72}$

Step 3   Divide.

$$
\begin{array}{r}
2 \\
3\overline{)72} \\
-6 \\
\hline
1
\end{array}
$$

← Multiply: $3 \times 2 = 6$
← Subtract: $7 - 6 = 1$

Step 4   Bring down the 2 ones.

$$
\begin{array}{r}
2 \\
3\overline{)72} \\
-6\downarrow \\
\hline
12
\end{array}
$$

Step 5   Divide.

$$
\begin{array}{r}
24 \\
3\overline{)72} \\
-6\downarrow \\
\hline
12 \\
-12 \\
\hline
0
\end{array}
$$

← Multiply: $3 \times 4 = 12$
← Subtract: $12 - 12 = 0$

Step 6   Use multiplication to check the quotient.

$3 \times 24 = 72$ ← This matches the dividend.

The quotient is correct.

Solution   **There will be 24 baseball cards in each case.**

When dividing greater numbers, first look at the digit in the greatest place of the dividend. Sometimes you will need to look at the first two places of the dividend.

Example 3

Divide.

$6\overline{)456}$

Strategy   **Divide each place from left to right.**

Step 1   Look at the digit in the greatest place of the dividend.

The digit is 4.

Because $4 < 6$, look at the digits in the first two places of the dividend.

$6\overline{)456}$

Step 2   Divide 45 tens by 6.

$$
\begin{array}{r}
7 \\
6\overline{)456} \\
-42 \\
\hline
3
\end{array}
$$
  ←— Multiply: $6 \times 7 = 42$
  ←— Subtract: $45 - 42 = 3$

Step 3   Bring down the ones. Divide 36 ones by 6.

$$
\begin{array}{r}
76 \\
6\overline{)456} \\
-42\downarrow \\
\hline
36 \\
-36 \\
\hline
0
\end{array}
$$
  ←— Multiply: $6 \times 6 = 36$
  ←— Subtract: $36 - 36 = 0$

Step 4   Use multiplication to check the quotient.

$6 \times 76 = 456$ ←— This matches the dividend.
The quotient is correct.

Solution $456 \div 6 = 76$

Example 4

Will raised $648 and Dara raised $522 for their favorite charities. They combined their money and donated the same amount to each of 3 charities. How much did each charity receive?

Strategy   **Decide how to solve the problem. Find the total amount of money raised. Then divide the sum by 3.**

Step 1   Add to find the total amount of money raised.

$648 + $522 = $1,170

Step 2   Divide to find how much each charity received.

$1,170 ÷ 3 = ☐

Divide each place from left to right.

```
      390
  3)1,170
   −9↓│
     27│
    −27↓
      00
     −00
       0
```

Step 3   Use multiplication to check the quotient.

3 × $390 = $1,170

Solution   **Each charity received $390.**

Coached Example

The circus has 8 equal sections of seats. There are 8,240 seats in all.

How many seats, _s_, are in each section?

Write a division sentence for the problem.

There are _____ seats in all.

There are _____ equal sections.

There are _s_ seats in each section.

_____ ÷ _____ = _____

Write the problem another way.

Divide each place from left to right.

$$8)\overline{8,240}$$

So, 8,240 ÷ 8 = _____.

There are _____ seats in each section.

Lesson Practice

Choose the correct answer.

1. Divide.

    $57 \div 3 = \boxed{}$

    A. 18
    B. 19
    C. 20
    D. 21

2. Gerald spent $81 on 3 tickets to a game. What is the cost of each ticket?

    A. $27
    B. $28
    C. $37
    D. $38

3. Divide.

    $65 \div 5 = \boxed{}$

    A. 10
    B. 11
    C. 12
    D. 13

4. An oak table cost $96. That is 8 times as much as a pine table costs. How much does a pine table cost?

    A. $12
    B. $11
    C. $10
    D. $9

5. Peter travels a total of 105 miles to and from work each week. He works 5 days each week. How many miles does Peter travel each day?

    A. 525 miles
    B. 210 miles
    C. 21 miles
    D. 15 miles

6. Divide.

    $264 \div 6 = \boxed{}$

    A. 44
    B. 46
    C. 47
    D. 48

7. Juanita bought a computer for $2,205. She will make 9 equal payments to pay off the computer. How much is each payment?

   A. $225

   B. $245

   C. $255

   D. $275

8. The workers paved 1,353 feet of a road today. That is 3 times as long as they paved yesterday. How much of the road did the workers pave yesterday?

   A.   450 feet

   B.   451 feet

   C.   551 feet

   D. 4,059 feet

9. Andrew, Nichole, and Sean are planning a canoe trip. They will split the cost of the supplies evenly. The table shows the supplies and the cost of each item.

Canoe Trip Supplies

Item	Cost
Canoe	$729
Life Jackets	$354
Paddles	$45

   A. How much will the supplies cost in all? Show your work.

   B. How much will each person pay for the canoe trip? Show your work.

Common Core State Standards:
4.NBT.6, 4.OA.3

Division with Remainders

Getting the Idea

The **remainder** is a number that is left after division has been completed.

You can write the remainder with the letter R. A remainder must be less than the divisor.

To check an answer with a remainder, first multiply the divisor by the quotient.

Then add that product to the remainder.

Look at the following example.

$13 \div 3 = 4 \text{ R}1$

The remainder 1 is less than the divisor 3.

To check the answer, multiply the divisor 3 by the quotient 4. $3 \times 4 = 12$

Then add the product 12 to the remainder 1. $12 + 1 = 13$

Example 1

Divide.

$61 \div 7 = \boxed{}$

Strategy **Divide each place from left to right.**

Step 1 Write the problem another way. Divide.

$$
\begin{array}{r}
8 \\
7\overline{)61} \\
-56 \\
\hline
5
\end{array}
$$

← Multiply $7 \times 8 = 56$

← Subtract $61 - 56 = 5$

There are 5 left over. This is the remainder.

Step 2 Write the remainder.

$$
\begin{array}{r}
8\ \textbf{R5} \\
7\overline{)61} \\
-56 \\
\hline
5
\end{array}
$$

Step 3   Check the answer.

Multiply the divisor by the quotient. Then add the remainder.

$(7 \times 8) + 5 =$

$56 \quad + 5 = 61$ ← This matches the dividend.
The answer is correct.

Solution $61 \div 7 = 8\ R5$

Example 2

Divide.

$531 \div 4 = \boxed{}$

Strategy   **Divide each place from left to right.**

Step 1   Set up the division. Start by dividing 5 hundreds by 4.

$$
\begin{array}{r}
1 \\
4\overline{)531} \\
-4 \\
\hline
1
\end{array}
$$
← Multiply $4 \times 1 = 4$
← Subtract $5 - 4 = 1$

Step 2   Continue dividing and write the remainder.

$$
\begin{array}{r}
132\ \mathbf{R3} \\
4\overline{)531} \\
-4 \\
\hline
13 \\
-12 \\
\hline
11 \\
-8 \\
\hline
3
\end{array}
$$

Step 3   Check the quotient.

$(4 \times 132) + 3 =$

$528 \quad + 3 = 531$ ← This matches the dividend.
The answer is correct.

Solution $531 \div 4 = 132\ R3$

In a division word problem with remainders, you may need to interpret the remainder. There are three ways to interpret the remainder.

1. Drop the remainder.
2. The remainder is the answer.
3. Add 1 to the quotient.

Example 3

A group of 124 chorus members are going to a concert. A van can take 9 members. How many vans are needed to get all of the members to the concert?

Strategy   **Divide. Then interpret the remainder.**

Step 1   Write the division sentence for the problem.

There are 124 members. Each van can take 9 members.

Let v represent the number of vans needed.

$124 \div 9 = v$

Step 2   Divide each place from left to right. Write the remainder.

$$
\begin{array}{r}
13\ \text{R}7 \\
9\overline{)124} \\
-9\downarrow \\
\hline
34 \\
-27 \\
\hline
7
\end{array}
$$

Step 3   Check the answer.

$(9 \times 13) + 7 =$

$117 \quad + 7 = 124$ ← This matches the dividend.
                                    The answer is correct.

Step 4   Interpret the remainder.

The answer, 13 R7, means 13 full vans with 7 members left over. Those 7 members remaining need to be driven. So one more van is needed. Add 1 to the quotient.

$13 + 1 = 14$

Solution   **To get all the members to the concert, 14 vans are needed.**

Example 4

A factory made 2,285 tea candles to be shipped to 8 different stores. Each store will receive the same number of candles with some left over. How many candles will each store receive?

Strategy **Divide. Then interpret the remainder.**

Step 1 Write the division sentence for the problem.

There are 2,285 candles to be shipped to 8 stores.

Let c represent the number of candles each store will receive.

$2,285 \div 8 = c$

Step 2 Set up the division. Divide each place from left to right.

$$
\begin{array}{r}
285 \text{ R5} \\
8\overline{)2285} \\
-16 \\
\hline
68 \\
-64 \\
\hline
45 \\
-40 \\
\hline
5
\end{array}
$$

Step 3 Check the quotient.

$(8 \times 285) + 5 =$

$2,280 \quad + 5 = 2,285$ ⟵ This matches the dividend. The quotient is correct.

Step 4 Interpret the remainder.

The answer, 285 R5, means 285 candles for each store with 5 candles left over. Since the question asks for the number of candles each store will receive, drop the remainder.

Solution **Each store will receive 285 candles.**

Coached Example

Nita has a 250-inch roll of ribbon. She needs to cut as many 9-inch pieces as she can from the roll. How many 9-inch pieces can Nita cut? What is the length of the ribbon left over?

Write the division sentence for the problem.

Nita has _____ inches of ribbon.

Each piece she will cut is _____ inches long.

Let p represent _____.

_____ ÷ _____ = p

Set up the division. Divide each place from left to right.

9)‾250‾

Check the quotient.

Multiply the divisor by the quotient. Add that product to the remainder.

(9 × _____) + _____ = p

_____ + _____ = _____

Does that match the dividend? _____ Is your answer correct? _____

Interpret the remainder.

The answer, _____ R_____, means Nita can cut _____ 9-inch pieces with _____ inches left over.

Nita can cut _____ 9-inch pieces.

The length of the ribbon left over is _____ inches.

Lesson Practice

Choose the correct answer.

1. Divide.

   $$65 \div 3 = \boxed{}$$

   A. 20 R4

   B. 21 R2

   C. 22 R1

   D. 22 R2

2. Which division sentence has a remainder?

   A. $99 \div 3 = \boxed{}$

   B. $38 \div 2 = \boxed{}$

   C. $77 \div 6 = \boxed{}$

   D. $65 \div 5 = \boxed{}$

3. The fourth-grade class collected 58 bottles to recycle. The students will pack 8 bottles in each box. The recycling depot only accepts full boxes. How many boxes will the class take to the recycling depot?

   A. 3

   B. 6

   C. 7

   D. 9

4. Divide.

   $$267 \div 6 = \boxed{}$$

   A. 44 R3

   B. 46 R1

   C. 47 R3

   D. 48 R7

5. Divide.

   $$988 \div 7 = \boxed{}$$

   A. 142 R6

   B. 142 R1

   C. 141 R6

   D. 141 R1

6. A total of 113 people signed up to take tennis lessons at the recreation center. Five people can have lessons every hour. How many hours are needed for everyone to have a lesson?

   A. 20 hours

   B. 22 hours

   C. 23 hours

   D. 25 hours

7. Divide.

$$2{,}603 \div 9 = \boxed{}$$

A. 289 R2

B. 289 R6

C. 290 R2

D. 291 R2

8. Nicole found that $3{,}429 \div 4 = 857$ R1. Which can she use to check that her answer is correct?

A. $(1 \times 3{,}429) + 4$

B. $(4 \times 857) + 1$

C. $(4 \times 1) + 857$

D. $(4 \times 857) \times 1$

9. The school fair committee needs enough pizzas for 275 people. Each pizza has 8 slices. How many pizzas does the committee need to order so that each person can have 1 slice?

A. Write a number sentence for the problem. Use a variable to represent the number of pizzas.

_____

B. How many pizzas does the committee need to order so that each person can have 1 slice? Show your work. Explain your answer.

_____

_____

_____

Common Core State Standards:
4.NBT.5, 4.NBT.6

Multiply and Divide by Multiples of 10, 100, and 1,000

Getting the Idea

A **multiple** of 10 is any counting number multiplied by 10.
A multiple of 100 is any counting number multiplied by 100.
A multiple of 1,000 is any counting number multiplied by 1,000.

To multiply a number by a multiple of 10, 100, or 1,000, multiply the number by the nonzero digit of the multiple of 10, and put one, two, or three zeros at the end of the product.

$$8 \times 20 = 160$$
$$8 \times 200 = 1,600$$
$$8 \times 2,000 = 16,000$$

You can use mental math to multiply a number by multiples of 10, 100, and 1,000.

Example 1

What is 2×90?

Strategy   **Use mental math.**

Step 1   Multiply 2 by the nonzero digit of the multiple of 10.

9 is the nonzero digit of the multiple of 10. $2 \times 9 = 18$

Step 2   There is one zero in 90.

Put one zero at the end of the product: 18**0**

Solution $2 \times 90 = 180$

Example 2

A netbook computer costs $300. A desktop computer costs 5 times as much as a netbook computer. How much does the desktop computer cost?

Strategy   **Use mental math.**

Step 1   Write a multiplication sentence for the problem.

The netbook costs $300.

The desktop costs 5 times as much.

Let n represent the cost of the desktop.

$300 \times 5 = n$

Step 2   Multiply 5 by the nonzero digit of the multiple of 100.

3 is the nonzero digit of the multiple of 100. $5 \times 3 = 15$

Step 3   There are 2 zeros in 300.

Put two zeros at the end of the product: 15**00**

So, $5 \times 300 = 1,500$.

Solution   **The desktop computer costs $1,500.**

Example 3

A store ordered 6 boxes of gumballs. Each box has 4,000 gumballs. How many gumballs in all did the store order?

Strategy   **Use mental math.**

Step 1   Write the multiplication sentence for the problem.

There are 6 boxes.

Each box has 4,000 gumballs.

Let g represent the total number of gumballs.

$6 \times 4,000 = g$

Step 2   Multiply 6 by the nonzero digit of the multiple of 1,000.

$6 \times 4 = 24$

Step 3   There are 3 zeros in 4,000.

Put 3 zeros at the end of the product: 24,**000**

So, $6 \times 4,000 = 24,000$.

Solution   **The store ordered 24,000 gumballs in all.**

Dividing a number by 10 is the opposite of multiplying by 10.

Instead of putting a zero at the end of a number, you take away a zero.

For example, $140 \div 10 = 14$.

To divide a number by 100, take away two zeros from the dividend.

For example, $1,400 \div 100 = 14$.

Example 4

Tanya has collected 90 dimes in a jar. She wrapped the dimes in rolls of 10 to bring to the bank. How many rolls of dimes did Tanya wrap?

Strategy   **Use mental math.**

Step 1   Write the division sentence for the problem.

There are 90 dimes.

There are 10 dimes in each roll.

Let r represent the number of rolls.

$90 \div 10 = r$

Step 2   Divide.

The divisor is 10, so take away a zero from the dividend.

$90 \div 10 = 9$

Solution   **Tanya wrapped 9 rolls of dimes.**

Coached Example

Mr. Cassidy typed 8,200 words in a report. He can type 100 words a minute. How many minutes did it take Mr. Cassidy to type his report?

Write the division sentence for the problem.

He typed _____ words in a report.

He can type _____ words a minute.

Let m represent the number of _____ it took to type the report.

_____ ÷ _____ = _____

Use mental math.

The divisor is _____, so take away _____ zeros from the dividend.

$8,200 \div 100 = $ _____

It took Mr. Cassidy _____ minutes to type his report.

Lesson Practice

Choose the correct answer.

1. Multiply.

$$6 \times 30 = \square$$

- **A.** 18
- **B.** 180
- **C.** 360
- **D.** 1,800

2. Jackie rides her bicycle 9 miles a day. If she does this for 10 days, how many miles will she ride in all?

- **A.** 9 miles
- **B.** 90 miles
- **C.** 900 miles
- **D.** 9,000 miles

3. Multiply.

$$2 \times 700 = \square$$

- **A.** 14
- **B.** 140
- **C.** 1,400
- **D.** 14,000

4. Samuel is buying shirts for the soccer team. The shirts cost $8 each. He ordered 100 shirts. How much did Samuel pay in all for the shirts?

- **A.** $8
- **B.** $80
- **C.** $800
- **D.** $8,000

5. Multiply.

$$3 \times 8,000 = \square$$

- **A.** 24
- **B.** 240
- **C.** 2,400
- **D.** 24,000

6. Divide.

$$1,000 \div 10 = \square$$

- **A.** 1,000
- **B.** 100
- **C.** 10
- **D.** 1

7. There were 5,000 entries in a contest. One in every 100 entries will win a prize. How many prizes will be given?

A. 5,000

B. 500

C. 50

D. 5

8. Divide.

$$1,300 \div 100 = \boxed{}$$

A. 1,300

B. 130

C. 13

D. 3

9. A standard room at the Copperfield Hotel costs $100 a night. A suite at the Copperfield Hotel costs $200 a night.

A. The Browns stayed in a suite for 4 nights. How much did the Browns pay for their stay at the Copperfield Hotel? Explain how you solved the problem.

_____

_____

_____

B. Vicki paid $700 for her stay in a standard room. How many nights did Vicki stay at the Copperfield Hotel? Explain how you solved the problem.

_____

_____

_____

Domain 1: Cumulative Assessment for Lessons 1–10

1. Which sentence is true?

   A. 412,440 > 412,549

   B. 416,543 < 415,811

   C. 413,303 > 413,030

   D. 411,312 < 411,231

2. Which number has the digit 4 in the thousands place and the digit 8 in the hundreds place?

   A. 123,483

   B. 254,837

   C. 368,448

   D. 459,804

3. A green shirt costs $6. A yellow shirt costs 4 times as much as the green shirt. How much does the yellow shirt cost?

   A. $2

   B. $10

   C. $16

   D. $24

4. Multiply.

   $3{,}652 \times 4 = \boxed{}$

   A. 1,468

   B. 14,408

   C. 14,608

   D. 14,628

5. Which correctly shows the multiplicative identity property of 1?

   A. $32 \times 5 = 5 \times 32$

   B. $21 \times 1 = 21$

   C. $0 \times 22 = 0$

   D. $4 \times (31 \times 9) = (4 \times 31) \times 9$

6. Which shows another way to write this number sentence?

   $23 \times 9 = \boxed{}$

   A. $(20 + 9) \times (3 + 9)$

   B. $(20 + 9) + (3 + 9)$

   C. $(20 \times 9) + (3 \times 9)$

   D. $(20 \times 9) \times (3 \times 9)$

7. This week Doreen baked 48 muffins. That was 6 times as many muffins as she baked last week. How many muffins did Doreen bake last week?

   A. 5

   B. 6

   C. 7

   D. 8

8. Shelli is trading in 750 tickets that she won at the arcade for 6 toy bracelets. Each bracelet costs the same number of tickets. How many tickets does each bracelet cost?

   A. 125

   B. 744

   C. 756

   D. 4,500

9. Divide.

   $$1,900 \div 100 = \boxed{}$$

10. Willie has 394 trading cards. He wants to put the cards in 4 albums. Each album will have the same number of cards. He will give the extra cards to his brother. How many cards will be in each album?

    A. Write a number sentence for the problem. Use a variable to represent the quotient.

    _____

    B. Solve the number sentence you wrote for Part A. Explain your answer.

    _____

    _____

Domain 2

Operations and Algebraic Thinking

Domain 2: Diagnostic Assessment for Lessons 11–17

Domain 2: Cumulative Assessment for Lessons 11–17

Domain 2: Diagnostic Assessment for Lessons 11–17

1. Which is a factor pair of 36?

    A. {4, 8}

    B. {4, 12}

    C. {5, 7}

    D. {6, 6}

2. At a snow cone stand, 2,843 snow cones were made on Friday and 3,802 snow cones were made on Saturday. How many snow cones were made on Friday and Saturday in all?

    A. 6,645

    B. 6,641

    C. 5,645

    D. 959

3. Which number is **not** a multiple of 6?

    A. 18   C. 43

    B. 30   D. 54

4. Caroline is watching a movie that is 207 minutes long. To the nearest hundred minutes, what is the length of the movie?

    A. 200 minutes

    B. 210 minutes

    C. 220 minutes

    D. 310 minutes

5. A store's grand opening day had 5,382 customers. To the nearest thousand, about how many customers did the store have on its grand opening day?

    A. 5,400   C. 5,300

    B. 5,380   D. 5,000

6. Students read books for a fundraiser. The table shows the number of books read by the students at four different schools.

    Books Read

School	Number of Books
Lincoln	5,130
Smith	3,205
Cherry Hill	6,089
Scotchtown	4,967

    To the nearest thousand, what is the total number of books read?

    A. 16,000   C. 19,000

    B. 18,000   D. 21,000

7. Which is the best way to estimate 37 × 41?

    A. 30 × 40

    B. 40 × 40

    C. 40 × 50

    D. 30 × 50

8. Jake has 87 model cars to display on 7 shelves. He will put about the same number of cars on each shelf. Which is the best estimate for the number of model cars on each shelf?

   A. 10

   B. 12

   C. 14

   D. 15

9. What is next number of this pattern?

   214 225 236 247 258 <u> ? </u>

_____

10. The table shows the number of points that 4 players scored in a game.

Player Scores

Player	Number of Points
Jill	3,423
Manny	2,875
Eric	4,148
Karen	3,039

   A. How many points did Jill and Karen score in all? Show your work.

   B. How many more points did Eric score than Manny? Show your work.

Factors and Multiples

Common Core State Standard:
4.OA.4

Getting the Idea

Factors are numbers that are multiplied together to get a product.

Every whole number greater than 1 has at least one pair of factors: 1 and itself.

For example, 1 and 10 is a factor pair of 10. Another factor pair of 10 is 2 and 5.

You can use an area model to find factor pairs.

Example 1

What are the factor pairs of 8?

Strategy   **Use area models.**

Step 1   Draw an area model that has 8 squares.

The area model shows 1 square by 8 squares.

One factor pair is 1 and 8.

Step 2   Draw another area model that has 8 squares.

The area model shows 2 squares by 4 squares.

Another factor pair is 2 and 4.

Step 3   You cannot make a different area model that has 8 squares.

List the factor pairs from the two area models.

Solution   **There are two factor pairs of 8: 1 and 8, 2 and 4.**

You can write a factor pair using braces. For example, one factor pair of 14 is {2, 7}.

Example 2

List the factor pairs of 24.

Strategy   **Use a multiplication table.**

Step 1   Write the first factor pair of 24.

Every whole number greater than 1 has 1 and itself as factors.

$1 \times 24 = 24$

So, 1 and 24 is one factor pair.

Step 2   Find all the 24s inside the multiplication table.

×	0	1	2	3	4	5	6	7	8	9	10	11	12
1	0	1	2	3	4	5	6	7	8	9	10	11	12
2	0	2	4	6	8	10	12	14	16	18	20	22	24
3	0	3	6	9	12	15	18	21	24	27	30	33	36
4	0	4	8	12	16	20	24	28	32	36	40	44	48
5	0	5	10	15	20	25	30	35	40	45	50	55	60
6	0	6	12	18	24	30	36	42	48	54	60	66	72
7	0	7	14	21	28	35	42	49	56	63	70	77	84
8	0	8	16	24	32	40	48	56	64	72	80	88	96
9	0	9	18	27	36	45	54	63	72	81	90	99	108
10	0	10	20	30	40	50	60	70	80	90	100	110	120
11	0	11	22	33	44	55	66	77	88	99	110	121	132
12	0	12	24	36	48	60	72	84	96	108	120	132	144

Step 3   Write a number sentence for each factor pair of 24.

$2 \times 12 = 24$

$3 \times 8 = 24$

$4 \times 6 = 24$

Some factors are used more than once.

You only need to list them once.

Step 4   List all the factor pairs of 24.

1 and 24

2 and 12

3 and 8

4 and 6

Solution   **The factor pairs of 24 are {1, 24}, {2, 12}, {3, 8}, and {4, 6}.**

A **multiple** is the product of two factors. Multiples form a skip-counting pattern. To find the first few multiples of a number, keep that number as one factor, and multiply by 1, then 2, then 3, and so on. Here are eight multiples of 5:

$5 \times 1 = \mathbf{5}$ $5 \times 2 = \mathbf{10}$ $5 \times 3 = \mathbf{15}$ $5 \times 4 = \mathbf{20}$

$5 \times 5 = \mathbf{25}$ $5 \times 6 = \mathbf{30}$ $5 \times 7 = \mathbf{35}$ $5 \times 8 = \mathbf{40}$

You can also use a multiplication table to find multiples of a number. Read down a column, or to the right along a row, to find multiples of a number.

You can use a square area model to show a multiple with two of the same factors, such as $5 \times 5 = 25$.

Example 3

Is 36 a multiple of 6?

Strategy   List multiples of 6.

$6 \times 1 = 6$

$6 \times 2 = 12$

$6 \times 3 = 18$

$6 \times 4 = 24$

$6 \times 5 = 30$

$6 \times 6 = 36$

Solution   Yes, 36 is a multiple of 6.

A **prime number** is a whole number that has exactly two factors, 1 and itself.
A **composite number** is a whole number that has more than one factor pair.
The number 1 is neither a prime number nor a composite number.

Example 4

Is 17 a prime number or a composite number?

Strategy   **Find the factor of 17.**

Step 1   Draw an area model that has 17 squares.

The area model shows 1 square by 17 squares.

Two factors of 17 are 1 and 17.

Step 2   Decide if you can make a different area model with 17 squares.

No, you cannot make another area model.

Because 17 has exactly two factors, it is a prime number.

Solution   **17 is a prime number.**

Example 5

Is 4 a prime number or a composite number?

Strategy   **Find the factors of 4.**

List the factors of 4.

The factors of 4 are 1, 2, and 4.

Solution   **Because 4 has more than two factors, it is a composite number.**

Coached Example

Paige has some dollar bills that she wants to exchange for quarters. She can exchange each dollar for 4 quarters. Can she get exactly 25 quarters by exchanging her dollar bills?

Decide whether 25 is a multiple of 4.

Use the pattern of multiples of 4.

$4 \times 1 =$ _____

$4 \times 2 =$ _____

$4 \times 3 =$ _____

$4 \times 4 =$ _____

$4 \times 5 =$ _____

$4 \times 6 =$ _____

$4 \times 7 =$ _____

The number 25 is between the products _____ and _____.

25 _____ a multiple of 4.

Paige _____ **get exactly 25 quarters by exchanging dollar bills.**

Lesson Practice

Choose the correct answer.

1. Which are **not** factors of 36?

   A. 2 and 16

   B. 3 and 12

   C. 4 and 9

   D. 6 and 6

2. Which number has 3 and 9 as factors?

   A. 12

   B. 24

   C. 27

   D. 49

3. Which number is a multiple of 4?

   A. 9

   B. 10

   C. 11

   D. 12

4. Dustin had 28 grapes. He put the same number of grapes in each bag. Which group of bags did Dustin make?

   A. 2 bags of 12 grapes

   B. 3 bags of 11 grapes

   C. 4 bags of 7 grapes

   D. 6 bags of 8 grapes

5. Kate's street address is a number with only two factors. Which mailbox could be Kate's?

   A.

   B.

   C.

   D.

6. Which is a multiple of 8?

   A. 73

   B. 48

   C. 23

   D. 18

7. Which number is **not** a prime number?

   A. 2

   B. 5

   C. 7

   D. 16

8. Which number is a prime number and a factor of 18?

   A. 3

   B. 6

   C. 9

   D. 18

9. What number between 48 and 58 is a prime number?

   _____

10. Ms. Henley wrote these numbers on the board.

    14   29   32   47   55   64

    A. Which numbers are prime numbers? Explain your answer.

       _____

       _____

    B. Which numbers are **not** prime numbers? Prove your answer by making a list of all the factor pairs of each number.

       _____

       _____

       _____

       _____

       _____

Add Whole Numbers

Common Core State Standards:
4.NBT.4, 4.OA.3

Getting the Idea

You can **add** to find the total when two or more groups are joined.

Here are the parts in an addition sentence.

2,411   +   3,524   =   5,935

addend   **addend**   **sum**

When you use paper and pencil to add, line up the digits on the ones place. Add the digits from right to left. If the sum of the digits in a column is 10 or greater, you will need to **regroup**.

Example 1

The table shows the number of miles Ms. Davis flew on each flight one day.

Miles Flown

From	To	Number of Miles
Boston, MA	Houston, TX	1,868
Houston, TX	Los Angeles, CA	1,524

How many miles did Ms. Davis fly in all?

Strategy   **Write an addition sentence, then solve.**

Step 1   Write the addition sentence for the problem.

Let m represent the total number of miles.

$1,868 + 1,524 = m$

Step 2   Line up the digits on the ones place. Add from right to left.

Add the ones: $8 + 4 = 12$.

Regroup 12 ones as 1 ten 2 ones.

$$
\begin{array}{r}
1 \\
1,868 \\
+\ 1,524 \\
\hline
2
\end{array}
$$

Step 3   Add the tens: $1 + 6 + 2 = 9$.

$$
\begin{array}{r}
1 \\
1,868 \\
+\ 1,524 \\
\hline
92
\end{array}
$$

Step 4   Add the hundreds: $8 + 5 = 13$.

Regroup 13 hundreds as 1 thousand 3 hundreds.

$$
\begin{array}{r}
1\ \ 1 \\
1,868 \\
+\ 1,524 \\
\hline
392
\end{array}
$$

Step 5   Add the thousands: $1 + 1 + 1 = 3$.

$$
\begin{array}{r}
1\ \ 1 \\
1,868 \\
+\ 1,524 \\
\hline
3,392
\end{array}
$$

Solution   **Ms. Davis flew 3,392 miles in all.**

Example 2

Add.

$$8,715 + 6,409 = \boxed{}$$

Strategy   **Line up the digits on the ones place. Add from right to left.**

$$
\begin{array}{r}
1\ \ 1 \\
8,715 \\
+\ 6,409 \\
\hline
15,124
\end{array}
$$

Solution   **$8,715 + 6,409 = 15,124$**

Example 3

The table shows the number of points needed to trade for prizes.

Prize Points Needed for Trade

Prize	Number of Points
$100 Gift Card	5,950
MP3 Player	10,135
DVD Player	5,885

Alex has 13,050 points. For which two prizes can Alex trade his points?

Strategy **Find the sum of the points for two prizes. Then compare the sum to Alex's points.**

Step 1 Find the sum for the $100 gift card and MP3 player.

$$
\begin{array}{r}
1 \\
5,950 \\
+\ 10,135 \\
\hline
16,085
\end{array}
$$

Alex does not have enough points because 16,085 > 13,050.

Step 2 Find the sum for the $100 gift card and DVD player.

$$
\begin{array}{r}
1\ 1 \\
5,950 \\
+\ 5,885 \\
\hline
11,835
\end{array}
$$

Alex does have enough points because 11,835 < 13,050.

Step 3 Find the sum for the MP3 player and DVD player.

$$
\begin{array}{r}
1\ 11 \\
10,135 \\
+\ 5,885 \\
\hline
16,020
\end{array}
$$

Alex does not have enough points because 16,020 > 13,050.

Solution **Alex can trade his points for the gift card and the DVD player.**

Coached Example

Last year, a club had 12,468 members. This year, the club has 8,271 more members. How many members are in the club this year?

Write the addition sentence for the problem.

The club had _____ members last year.

The club has _____ more members this year.

Let *m* represent the total number of members in the club this year.

_____ + _____ = _____

Set up the problem.

Line up the digits on the ones place.

Add each place from right to left. Regroup if necessary.

The sum is _____.

There are _____ members in the club this year.

Lesson Practice

Choose the correct answer.

1. Add.

$$3,674$$
$$+\ 4,369$$

   A. 7,933

   B. 7,943

   C. 8,033

   D. 8,043

2. Add.

$$6,000 + 3,173 = \boxed{}$$

   A. 2,827

   B. 9,173

   C. 9,827

   D. 10,937

3. The High Bridge in Kentucky is 1,125 feet long. The Brooklyn Bridge in New York is 4,864 feet longer than the High Bridge. What is the length of the Brooklyn Bridge?

   A. 3,739 feet

   B. 5,864 feet

   C. 5,989 feet

   D. 6,989 feet

4. Add.

$$5,215 + 3,107 = \boxed{}$$

   A. 8,302

   B. 8,312

   C. 8,321

   D. 8,322

5. Add.

$$43,674$$
$$+\ 3,372$$

   A. 46,046

   B. 46,946

   C. 47,046

   D. 77,396

6. At a tire factory, 3,166 tires were made on Thursday and 2,941 tires were made on Friday. How many tires were made on Thursday and Friday in all?

   A. 6,107

   B. 6,557

   C. 6,602

   D. 6,972

7. Mr. Newton ate 2,245 calories yesterday. He ate 2,583 calories today. How many calories did Mr. Newton eat in all?

   A. 4,742

   B. 4,828

   D. 4,928

   C. 5,828

8. Two years ago, Ms. Bolton bought a used car that showed 14,854 miles on the odometer. Last year, she drove 8,240 miles. This year, she drove 9,273 miles. How many miles does the odometer show now?

   A. 17,513 miles

   B. 24,127 miles

   C. 31,367 miles

   D. 32,367 miles

9. A recycling center recycles plastic bottles, aluminum cans, and glass bottles. The table shows the number of each material the center recycled in one day.

Materials Recycled

Material	Number Recycled
Plastic bottles	12,847
Aluminum cans	9,659
Glass bottles	3,273

A. Did the center recycle more plastic bottles or more aluminum cans and glass bottles combined that day? Show your work.

B. How many materials in all did the center recycle that day? Show your work.

Subtract Whole Numbers

Common Core State Standards:
4.NBT.4, 4.OA.3

Getting the Idea

You can **subtract** to find how many are left when you take something away.

Here are the parts in a subtraction sentence.

$$3,667 \quad - \quad 1,243 \quad = \quad 2,424$$

minuend   **subtrahend**   **difference**

When you use paper and pencil to subtract, remember to line up the digits on the ones place. Subtract each digit from right to left. If the digit in the minuend is less than the digit in the subtrahend, you have to regroup.

Example 1

Peter scored 5,189 points playing a video game. Jacob scored 1,778 points playing the same game. How many more points did Peter score than Jacob?

Strategy   **Write a subtraction sentence, then solve.**

Step 1   Write the subtraction sentence for the problem.

         Peter scored   5,189 points.

         Jacob scored 1,778 points.

         Let p represent how many more points Peter scored than Jacob.

         $5,189 - 1,778 = p$

Step 2   Line up the digits on the ones place. Subtract from right to left.

         Subtract the ones: $9 - 8 = 1$

$$\begin{array}{r} 5,18\mathbf{9} \\ -\ 1,77\mathbf{8} \\ \hline \mathbf{1} \end{array}$$

Step 3   Subtract the tens: $8 - 7 = 1$

$$\begin{array}{r} 5,1\mathbf{8}9 \\ -\ 1,7\mathbf{7}8 \\ \hline \mathbf{11} \end{array}$$

Step 4 There are not enough hundreds to subtract.

Regroup 1 thousand as 10 hundreds. Now there are 11 hundreds.

$$
\begin{array}{r}
{}^{4}{}^{11} \\
\cancel{5},\cancel{1}\,8\,9 \\
-\;1{,}7\,7\,8 \\
\hline
1\,1
\end{array}
$$

Step 5 Subtract the hundreds: $11 - 7 = 4$

$$
\begin{array}{r}
{}^{4}{}^{11} \\
\cancel{5},\cancel{1}\,8\,9 \\
-\;1{,}\mathbf{7}\,7\,8 \\
\hline
\mathbf{4}\,1\,1
\end{array}
$$

Step 6 Subtract the thousands: $4 - 1 = 3$

$$
\begin{array}{r}
{}^{4}{}^{11} \\
\cancel{5},\cancel{1}\,8\,9 \\
-\;\mathbf{1}{,}7\,7\,8 \\
\hline
\mathbf{3}{,}4\,1\,1
\end{array}
$$

Solution   **Peter scored 3,411 more points than Jacob.**

When subtracting with zeros in the minuend, you may have to regroup from more than one place.

Example 2
Subtract.

$4{,}007 - 1{,}526 = \boxed{}$

Strategy   **Line up the digits on the ones place. Subtract from right to left.**

Step 1 Line up the digits on the ones place.

Subtract the ones: $7 - 6 = 1$

$$
\begin{array}{r}
4{,}00\mathbf{7} \\
-\;1{,}52\mathbf{6} \\
\hline
\mathbf{1}
\end{array}
$$

Step 2 There are not enough tens to subtract and no hundreds to regroup.

Regroup 1 thousand as 10 hundreds.

Then regroup 1 hundred as 10 tens.

$$
\begin{array}{r}
9 \\
3 \;\; \cancel{10} \;\; 10 \\
\cancel{4}, \cancel{0} \; \cancel{0} \; 7 \\
-\,1{,}5\,2\,6 \\
\hline
1
\end{array}
$$

Step 3 Subtract the tens: $10 - 2 = 8$

$$
\begin{array}{r}
9 \\
3 \;\; \cancel{10} \;\; 10 \\
\cancel{4}, \cancel{0} \; \cancel{0} \; 7 \\
-\,1{,}5\,2\,6 \\
\hline
8\;1
\end{array}
$$

Step 4 Subtract the hundreds: $9 - 5 = 4$

$$
\begin{array}{r}
9 \\
3 \;\; \cancel{10} \;\; 10 \\
\cancel{4}, \cancel{0} \; \cancel{0} \; 7 \\
-\,1{,}5\,2\,6 \\
\hline
4\;8\;1
\end{array}
$$

Step 5 Subtract the thousands: $3 - 1 = 2$

$$
\begin{array}{r}
9 \\
3 \;\; \cancel{10} \;\; 10 \\
\cancel{4}, \cancel{0} \; \cancel{0} \; 7 \\
-\,1{,}5\,2\,6 \\
\hline
2{,}4\,8\,1
\end{array}
$$

Solution $4{,}007 - 1{,}526 = 2{,}481$

Addition is the opposite, or the **inverse operation**, of subtraction.
You can check the answer to a subtraction problem using addition.

$$
\begin{array}{rr}
2{,}481 & 4{,}007 \\
+\,1{,}526 & -\,1{,}526 \\
\hline
4{,}007 & 2{,}481
\end{array}
$$

The sum is 4,007, which matches the minuend. So, the answer is correct.

Example 3

In a survey about favorite movies, 4,893 male students and 5,203 female students gave a response. There were also responses from 1,572 adults. How many more students than adults responded to the survey?

Strategy   **Write number sentences to represent the problem. Then solve.**

Step 1   Write an addition sentence to find the total number of students.

There were 4,893 male students.

There were 5,203 female students.

Let *s* represent the total number of students.

4,893 + 5,203 = *s*

Step 2   Find the total number of students.

Add. Regroup if necessary.

$$\begin{array}{r} 1 \\ 4,893 \\ +\ 5,203 \\ \hline 10,096 \end{array}$$

Step 3   Write a subtraction sentence to find how many more students than adults responded to the survey.

There were 10,096 students.

There were 1,572 adults.

Let *n* represent how many more students.

10,096 − 1,572 = *n*

Step 4   Find how many more students than adults responded to the survey.

Subtract. Regroup if necessary.

$$\begin{array}{r} 9 \\ 0\ 10\ 10 \\ \cancel{1}\cancel{0},\cancel{0}\ 9\ 6 \\ -\ \ 1,5\ 7\ 2 \\ \hline 8,5\ 2\ 4 \end{array}$$

Step 5   Use addition to check the subtraction.

$$
\begin{array}{r}
1 \\
8,524 \\
+ \ 1,572 \\
\hline
10,096
\end{array}
$$
    ← This matches the minuend.

The answer is correct.

Solution   **8,524 more students than adults responded to the survey.**

Coached Example

Lynn had $2,812 in her checking account. She spent $1,150 on a television and $665 on a video camera. How much does Lynn have left in her checking account?

Decide how to solve the problem.

_____ to find the total amount Lynn spent on the television and the video camera.

Then _____ the sum from the amount Lynn had in her checking account.

Add to find the total amount Lynn spent.

Add from right to left. Regroup if necessary.

Subtract to find how much Lynn has left in her checking account.

Subtract from right to left. Regroup if necessary.

Use addition to check the subtraction.

Lynn has $_____ left in her checking account.

Lesson Practice

Choose the correct answer.

1. Subtract.

    $$\begin{array}{r} 8{,}715 \\ -\ 5{,}923 \\ \hline \end{array}$$

    A. 2,792

    B. 2,802

    C. 2,892

    D. 3,792

2. Subtract.

    $6{,}000 - 2{,}173 = \boxed{}$

    A. 8,173

    B. 3,937

    C. 3,927

    D. 3,827

3. Sara scored 2,293 points in a video game. That was 1,536 points more than Elvin's score on the same game. How many points did Elvin score?

    A.   757

    B. 1,757

    C. 1,767

    D. 3,829

4. Subtract.

    $$\begin{array}{r} 9{,}534 \\ -\ 4{,}085 \\ \hline \end{array}$$

    A. 4,449   C. 5,549

    B. 5,449   D. 5,551

5. Subtract.

    $$\begin{array}{r} 18{,}143 \\ -\ 5{,}923 \\ \hline \end{array}$$

    A. 12,220

    B. 12,320

    C. 12,820

    D. 41,113

6. The highest point of the Great Smoky Mountains in Tennessee is 6,643 feet. The highest point of the mountain Jordan has climbed is 1,870 feet. How much higher is the highest point of the Great Smoky Mountains than the highest point of the mountain Jordan climbed?

    A. 4,663 feet

    B. 4,773 feet

    C. 4,873 feet

    D. 5,233 feet

7. There are 11,510 seats in the basketball arena. For one game, 9,465 seats were filled. How many seats were empty?

   A. 2,165

   B. 2,155

   C. 2,045

   D. 1,045

8. Carol bought a car with a final price of $17,067. This price included an interior leather package for $1,258 and satellite radio for $359. What was the price of the car without these features?

   A. $16,708

   B. $16,450

   C. $15,809

   D. $15,450

9. A toy company made $27,358 from selling game consoles. It also made $3,725 from video games and $8,440 from board games.

   A. How much more did the company make from board games than from video games? Show your work.

   B. How much more did the company make from game consoles than from video and board games combined? Explain how you solved the problem.

   _____

   _____

   _____

Common Core State Standard:
4.NBT.3

Round to the Nearest Ten, Hundred, or Thousand

Getting the Idea

You can **round** a number to the nearest 10 or 100. When rounding, you replace a number with one that tells *about* how much or *about* how many. Rounding gives a number close to the exact amount.

You can use a number line to help you round numbers.

A number line can help you decide which 10 or 100 a number is closer to.

Example 1

What is 128 rounded to the nearest 10?

Strategy   **Use a number line.**

   Step 1   Place 128 on a number line.

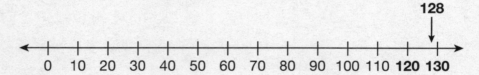

   Step 2   Decide whether 128 is closer to 120 or 130.

           128 is closer to 130 than to 120.

           128 rounds up to 130.

Solution   **128 rounded to the nearest 10 is 130.**

You can also use rounding rules to round numbers.

> **Rounding Rules**
>
> 1. Find the place you want to round to.
>
> 2. Look at the digit to the right of the place you are rounding to.
>
> 3. If the digit is 1, 2, 3, or 4, round down.
>   Leave the digit in the rounding place as is.
>
> 4. If the digit is 5, 6, 7, 8, or 9, round up.
>   Increase the digit in the rounding place by 1.
>
> 5. Change the digits to the right of the rounding place to zeros.

Example 2

A chandelier has 1,723 crystal pieces. To the nearest hundred, about how many crystal pieces does the chandelier have?

Strategy   **Use rounding rules to round to the nearest hundred.**

Step 1   Find the rounding place.

Underline the digit in the place you want to round to, the hundreds place.

1,7̲23

Step 2   Decide to round up or down.

Look at the digit to the right of the rounding place, in the tens place.

1,7̲23

The digit is 2. It is less than 5, so round down.

Step 3   Round 1,723 down to the nearest hundred.

Leave the digit in the hundreds place.

Change the digits to the right of the hundreds place to 0.

1,723 → 1,700

Solution   **To the nearest hundred, the chandelier has about 1,700 crystal pieces.**

Example 3

Quinn has 23,867 frequent flyer miles. To the nearest thousand, about how many frequent flyer miles does Quinn have?

Strategy **Use rounding rules to round to the nearest thousand.**

Step 1 Find the place you want to round to. Look at the digit to the right.

You are rounding to the nearest thousand, so find the thousands digit.

Then look at the hundreds digit.

23,867

Step 2 Decide to round up or down.

8 is greater than 5, so round up.

Step 3 Round 23,867 up to the nearest thousand.

Increase the thousands digit by 1.

Change the digits to the right of the thousands place to 0.

23,867 → 24,000

Solution **To the nearest thousand, Quinn has 24,000 frequent flyer miles.**

Example 4

Martin's Music Shop earned $12,445 in April, $15,125 in May, and $14,675 in June. During which two months did Martin's Music Shop earn about the same amount of money?

Strategy **Round to the nearest thousand. Then compare.**

Step 1 Round each money amount to the nearest thousand.

April: $12,445 rounds down to $12,000.

May: $15,125 rounds down to $15,000.

June: $14,675 rounds up to $15,000.

Step 2 Compare the amounts.

The amounts for May and June both rounded to $15,000.

Solution **Martin's Music Shop earned about the same amount of money in May and June.**

Coached Example

A game Web site received 129,354 hits in one day. To the nearest ten thousand, about how many hits did the game Web site receive that day?

The place to be rounded to is _____.

The digit in this place is _____.

The digit to the right of the rounding place is _____.

This digit is _____ than 5.

Since the digit to the right is greater than 5, round _____.

Change all the digits to the right of the rounding place to _____.

129,354 rounds to _____.

To the nearest ten thousand, the game Web site received _____ hits that day.

Lesson Practice

Choose the correct answer.

1. What is 33,719 rounded to the nearest thousand?

   A. 30,000

   B. 33,000

   C. 34,000

   D. 40,000

2. Which shows 144,683 rounded to the nearest ten thousand?

   A. 100,000

   B. 140,000

   C. 144,000

   D. 145,000

3. Which number does **not** round to 500?

   A. 459

   B. 486

   C. 521

   D. 550

4. Francesca has 281 marbles in her collection. To the nearest hundred, about how many marbles does Francesca have?

   A. 300

   B. 290

   C. 280

   D. 200

5. To the nearest ten, which number rounds to 320?

   A. 314

   B. 322

   C. 326

   D. 330

6. In the first week of August, 4,784 people visited a water park. To the nearest hundred, about how many people visited the water park in the first week?

   A. 4,800

   B. 4,700

   C. 4,600

   D. 4,000

7. A resort vacation for two people costs $1,577. To the nearest hundred dollars, about how much does the trip cost?

A. $2,000

B. $1,680

C. $1,600

D. $1,500

8. The table shows the number of books in each library.

Library	Number of Books
North Park	7,935
Central	8,162
Mountain View	7,493
Golden Hill	8,046

To the nearest thousand, which library has about 7,000 books?

A. North Park

B. Central

C. Mountain View

D. Golden Hill

9. For the county bake sale, the softball team baked 322 cookies, 295 brownies, and 248 muffins.

A. Round each type of baked good to the nearest hundred.

_____

_____

_____

B. The softball team baked about the same amount of two types of baked goods. What types were they? Explain your answer.

_____

_____

_____

Estimate Sums and Differences

Common Core State Standards:
4.OA.3

Getting the Idea

You can use rounded numbers to **estimate** sums and differences.

Your estimates will vary depending on what place you round to.

Rounding to tens	Rounding to hundreds	Rounding to thousands
4,252 → 4,250	4,252 → 4,300	4,252 → 4,000
+1,607 → +1,610	+1,607 → +1,600	+1,607 → +2,000
5,860	5,900	6,000

Example 1

Kendra scored 9,235 points in the first round of a computer game. She scored 8,790 points in the second round. About how many points did she score in all?

Strategy   **Round each number to its greatest place and then solve.**

> **Step 1**   Choose the operation.
>
> The words "in all" show that this is an addition problem.

> **Step 2**   Round each number to its greatest place: thousands.
>
> The number 9,235 rounds down to 9,000 because 2 < 5.
>
> The number 8,790 rounds up to 9,000 because 7 > 5.

> **Step 3**   Add the rounded numbers.
>
> $9,000 + 9,000 = 18,000$

Solution   **Kendra scored about 18,000 points in all.**

Example 2

Lee has 4,837 building blocks. He gave 1,175 building blocks to Sarah. To the nearest thousand, about how many blocks does Lee have left?

Strategy **Round each number to the thousands place and then solve.**

> **Step 1** Choose the operation.
>
>   "Gave" and "have left" shows that this is a subtraction problem.
>
> **Step 2** Round each number to the thousands place.
>
>   4,837 rounds up to 5,000 because 8 > 5.
>
>   1,175 rounds down to 1,000 because 1 < 5.
>
> **Step 3** Subtract the rounded numbers.
>
>   5,000 − 1,000 = 4,000

Solution **To the nearest thousand, Lee has about 4,000 building blocks left.**

When you need an exact answer, estimating before you add or subtract is a good way to see if your answer is reasonable. You can quickly tell if you made an error when adding or subtracting the numbers.

Example 3

Pam said that 37,249 + 46,867 is equal to 74,126. Is her answer reasonable?

Strategy **Estimate to check whether an answer is reasonable.**

> **Step 1** Estimate by rounding to thousands.
>
>   37,259 rounds down to 37,000 because 2 < 5.
>
>   46,867 rounds up to 47,000 because 8 > 5.
>
> **Step 2** Add the rounded numbers.
>
>   37,000 + 47,000 = 84,000
>
> **Step 3** Compare the estimate to the answer.
>
>   84,000 is not close to 74,216.

Solution **Pam's answer is not reasonable.**

Coached Example

A theme park had 74,868 visitors last month. This month, the park had 79,967 visitors. About how many more visitors did the theme park have this month than last month?

The word _____ tells me that I need an estimate.

The words "how many more" tell me that I need to _____ the numbers.

Estimate by rounding to thousands.

The number 74,868 rounds to _____.

The number 79,967 rounds to _____.

Subtract the rounded numbers.

_____ − _____ = _____

About _____ more people visited the theme park this month than last month.

Lesson Practice

Choose the correct answer.

1. Which is the best way to estimate 68 + 71?

   A. 60 + 70

   B. 60 + 80

   C. 70 + 70

   D. 80 + 70

2. Which is the best way to estimate 278 − 127?

   A. 200 − 100

   B. 300 − 100

   C. 200 − 200

   D. 300 − 200

3. Janelle cut 4 pieces of rope that measured 43 inches, 82 inches, 53 inches, and 27 inches. Which is the best estimate of the total length of all 4 pieces of rope?

   A. 100 inches

   B. 150 inches

   C. 200 inches

   D. 300 inches

4. A store had 135 customers in the morning and 884 customers in the afternoon. About how many more customers did the store have in the afternoon than in the morning?

   A. 600

   B. 800

   C. 900

   D. 1,000

5. A movie theater sold 23,874 tickets in August. It sold 15,345 tickets in September. About how many more tickets were sold in August than in September?

   A. 2,000

   B. 3,000

   C. 5,000

   D. 9,000

6. Reggie scored 3,219 points in the first round of a game. He scored 5,199 points in the second round. How many points did he score in all, to the nearest thousand?

   A. 2,000

   B. 4,000

   C. 8,000

   D. 9,000

Use the table for questions 7 and 8.

Distances Craig Biked

Year	Distance(in miles)
2006	1,287
2007	1,365
2008	2,082

7. To the nearest thousand, how much farther did Craig bike in 2008 than in 2007?

   A. 1,000 miles

   B. 2,000 miles

   C. 3,000 miles

   D. 4,000 miles

8. To the nearest hundred, how many miles did Craig bike in all?

   A. 4,000 miles

   B. 4,500 miles

   C. 4,700 miles

   D. 4,800 miles

9. Last month, Ms. Barkley spent $378 on food, $925 on rent, and $272 on utilities.

   A. What is the total amount Ms. Barkley spent on those three things last month? Show your work.

   B. Find the estimated total amount Ms. Barkley spent on those three things. Compare the estimate to the exact amount. Is your answer reasonable? Explain your answer.

   _____

   _____

   _____

Estimate Products and Quotients

Common Core State Standard:
4.OA.3

Getting the Idea

As with addition and subtraction problems, sometimes you do not need an exact answer to solve multiplication and division problems.

To estimate a product, round the numbers given in the problem. Then use mental math to find the estimated answer.

Example 1

Phil delivers 38 newspapers each Sunday. About how many Sunday newspapers does he deliver in a year? There are about 52 weeks in a year.

Strategy   **Round each number. Use mental math to multiply.**

Step 1   Decide how to solve the problem.

"About" tells you to estimate.

He delivers 38 newspapers each Sunday.

There are 52 weeks in a year.

Estimate 52 × 38.

Step 2   Round each number to the nearest 10.

$$52 \quad \times \quad 38 \quad = \quad \square$$
$$\downarrow \qquad\qquad \downarrow$$
$$50 \quad \times \quad 40 \quad = \quad \square$$

Step 3   Use mental math to multiply the rounded numbers.

Think: 5 × 4 = 20

50 × 40 = 2,000

Solution   **Phil delivers about 2,000 Sunday newspapers a year.**

You can use estimation to decide if an answer is reasonable. Before you find the exact answer, find the estimated product. Then compare the estimate to the exact answer to see if the answer makes sense.

Example 2

Ms. Barrows pays $479 to rent a car for a month. How much will Ms. Barrows pay to rent a car for 3 months?

Strategy   **Find the estimated answer. Then compare it to the exact answer.**

Step 1   Write the multiplication sentence for the problem.

She pays $479 for 1 month.

She rents the car for 3 months.

Let n represent the total amount she will pay for 3 months.

$3 \times \$479 = n$

Step 2   Estimate the answer.

$479 rounds up to $500.

$3 \times \$500 = \$1,500$

The answer should be about $1,500.

Step 3   Find the exact answer.

$$\begin{array}{r} 22 \\ \$479 \\ \times \quad 3 \\ \hline \$1437 \end{array}$$

Step 4   Compare the exact answer to the estimated answer.

$1,437 is close to $1,500.

The answer is reasonable.

Solution   **Ms. Barrows will pay $1,437 to rent a car for 3 months.**

You can use **compatible numbers** and mental math to estimate quotients. Compatible numbers are close to the exact numbers and are easy to compute with. To estimate $55 \div 7$, think of a number close to 55 that can be evenly divided by 7. 56 is close to 55 and $56 \div 7 = 8$. So, $55 \div 7$ is about 8.

Example 3

A basketball team scored 143 points in 3 games. If the team scored the same number of points in each game, about how many points did the team score in one game?

Strategy	Use compatible numbers to estimate the quotient.

Step 1 Decide how to solve the problem.

"About" tells you to estimate.

The team scored 143 points in 3 games.

Estimate $143 \div 3 = \square$.

Step 2 Find a number close to 143 that is compatible with 3.

150 is close to 143 and can be evenly divided by 3.

Step 3 Estimate the quotient.

$15 \div 3 = 5$, so $150 \div 3 = 50$.

Solution **The basketball team scored about 50 points a game.**

Sometimes you may need to change both numbers to find compatible numbers.

Example 4

Adele orders 135 roses for a reception. She wants to put the same number of roses in each vase. If she has 9 vases, how many roses will Adele put in each vase?

Strategy	Use compatible numbers to find the estimated answer. Then compare it to the exact answer.

Step 1 Write the division sentence for the problem.

She has 135 roses and 9 vases.

Find $135 \div 9 = \square$.

Step 2 Use compatible numbers to estimate the answer.

135 is close to 140. 9 is close to 10.

$140 \div 10 = 14$

The answer should be about 14.

Step 3 Find the exact answer.

$135 \div 9 = 15$

Step 4 Compare the exact answer to the estimated answer.

15 is close to 14. The answer is reasonable.

Solution **Adele will put 15 roses in each vase.**

Coached Example

A T-shirt factory shipped 7 boxes of shirts. Each box had 275 shirts. How many shirts did the factory ship in all?

Write the number sentence for the problem.

The words "in all" tell you to _____.

Find _____ × _____ = ☐.

Estimate the answer.

Round 275 to the nearest 100.

275 rounds to _____.

Multiply the rounded numbers.

_____ × _____ = _____

The answer should be about _____.

Find the exact answer.

Is the exact answer close to the estimated answer? _____

Is the answer reasonable? _____

The factory shipped _____ shirts in all.

Lesson Practice

Choose the correct answer.

1. Which is the best estimate for 1,589 × 4?

   A. 5,600

   B. 6,000

   C. 6,400

   D. 8,000

2. The local bakery makes 7 trays of oatmeal cookies each morning. Each tray holds 22 cookies. About how many cookies does the bakery make each morning?

   A. 140

   B. 180

   C. 210

   D. 300

3. Which is the best way to estimate 258 ÷ 5?

   A. 350 ÷ 5

   B. 300 ÷ 5

   C. 250 ÷ 5

   D. 200 ÷ 5

4. Which is the best way to estimate 314 ÷ 9?

   A. 270 ÷ 9

   B. 300 ÷ 9

   C. 314 ÷ 10

   D. 310 ÷ 10

5. Yesterday morning, a museum exhibit was open for 3 hours. There were 2,378 visitors each hour. Which is the best estimate for how many visitors in all 3 hours?

   A. 7,000

   B. 7,200

   C. 7,500

   D. 8,000

6. Which is the best estimate for 1,639 ÷ 4?

   A. 100

   B. 200

   C. 300

   D. 400

7. There are 6 bins of aluminum cans to be recycled. Each bin has 858 cans. Which is the best estimate for the number of cans in all 6 bins?

   A. 4,800

   B. 5,400

   C. 6,000

   D. 8,000

8. Crystal has taken 8 quizzes with a total score of 619. She received about the same score for each quiz. Which is the best estimate of the score for each quiz?

   A. 72

   B. 75

   C. 78

   D. 82

9. The fourth-grade class performed 4 shows for a total of 788 people.

   A. If each show had the same number of people, how many people attended each show? Show your work.

   B. Find the estimated quotient. Compare the estimate to the exact answer in Part A. Is your answer reasonable? Explain.

   _____

   _____

   _____

Patterns

Common Core State Standard:
4.0A.5

Getting the Idea

A **pattern** is a series of numbers or figures that follows a **rule**. In a number pattern, the numbers are the **terms**. The rule describes how each term is related to the next term.

This number pattern has 5 terms. The rule of the pattern is "add 4."

    27   31   35   39   43

You can find the next term in a number pattern by finding the rule.
Some patterns increase, so try adding or multiplying by the same number.
Some patterns decrease, so try subtracting or dividing by the same number.

Example 1

What is the next term of this pattern?

    15   22   29   36   __?__

Strategy   **Find the rule of the pattern.**

Step 1   Decide if the terms increase or decrease.

    The numbers increase.

Step 2   Find how many are between each term.

    If the numbers increase, use addition or multiplication.

    Try addition.

    $15 + ? = 22$ → $15 + \mathbf{7} = 22$

    $22 + ? = 29$ → $22 + \mathbf{7} = 29$

    $29 + ? = 36$ → $29 + \mathbf{7} = 36$

    Each number is 7 more than the previous number.

Step 3   Find a rule.

    A rule is "add 7."

Step 4   Use the rule to find the next term.

    Add 7 to the last number.

    $36 + 7 = 43$

Solution   **The next term of this pattern is 43.**

Notice another pattern of the numbers in Example 1.

15	22	29	36	42
odd	even	odd	even	odd

The terms in the pattern alternate between odd and even numbers.

The rule of the pattern is "add 7" and 7 is an odd number.

This rule will create alternating odd and even numbers because:

odd number + odd number = even number

even number + odd number = odd number

Example 2

The table shows how much money Martin had in his school lunch account at the end of each week for four weeks.

Week 1	Week 2	Week 3	Week 4	Week 5
$150	$135	$120	$105	?

If the pattern continues, how much money will Martin have in his account at the end of Week 5?

Strategy   **Find the rule of the pattern.**

Step 1   Decide if the terms increase or decrease.

       The numbers decrease.

Step 2   Find the rule.

       Since the numbers decrease, use subtraction or division.

       Try subtraction.

$$150 - ? = 135 \longrightarrow 150 - \mathbf{15} = 135$$
$$135 - ? = 120 \longrightarrow 135 - \mathbf{15} = 120$$
$$120 - ? = 105 \longrightarrow 120 - \mathbf{15} = 105$$

       The rule is "subtract 15."

Step 3   Use the rule to find the next term.

       Subtract 15 from the last number.

$$105 - 15 = 90$$

Solution   **If the pattern continues, Martin will have $90 in his account at the end of Week 5.**

Notice that the terms in the pattern in Example 2 also alternate between even and odd numbers. Remember that:

even number − odd number = odd number

odd number − odd number = even number

Example 3

Jenny is making a number pattern with 6 terms. The first term is 2.
The rule of the pattern is "multiply by 2." What are the six terms in Jenny's pattern?

Strategy   **Use the rule.**

Step 1   Identify the information given in the problem.

The pattern has 6 terms.

The pattern starts with 2.

The rule is "multiply by 2."

2   _?_   _?_   _?_   _?_   _?_

Step 2   Use the rule to find the next term of the pattern.

Multiply the first term, 2, by 2.

$2 \times 2 = 4$ ←   second term

Step 3   Use the rule to find the rest of the terms in the pattern.

$4 \times 2 = $**8**   ←   third term

$8 \times 2 = $**16**   ←   fourth term

$16 \times 2 = $**32**   ←   fifth term

$32 \times 2 = $**64**   ←   sixth term

Notice that the terms are all even numbers.

The rule of the pattern is multiply by 2, so each term doubles the previous term.

Solution   **The six terms in Jenny's pattern are 2, 4, 8, 16, 32, and 64.**

A pattern made up of figures also uses a rule. Use the rule to continue the pattern. You can use a table to help you.

Example 4

What is the 17th figure in this pattern?

Strategy   **Use a table.**

Step 1   Find the rule of the pattern.

This is a repeating pattern.

1 triangle, 1 rectangle, 2 circles

Step 2   Use the table to extend the pattern.

You know the first 8 figures.

Make a table for figures 9 through 17.

Figure	9	10	11	12	13	14	15	16	17
Shape	△	▯	○	○	△	▯	○	○	△

Solution   **The 17th figure is a triangle.**

Example 5

How many dots are in the 5th figure?

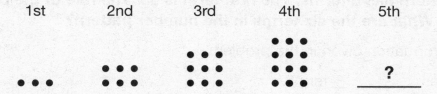

| 1st | 2nd | 3rd | 4th | 5th |

?
_____

Strategy   **Find the rule of the pattern. Use a table.**

Step 1   Count the number of dots in each figure.

Figure	1st	2nd	3rd	4th
Number of Dots	3	6	9	12

Step 2   Find the rule.

Each figure has 1 row of 3 dots more than the previous figure.

Figure	1st	2nd	3rd	4th
Number of Dots	3	6	9	12
	1×3	2×3	3×3	4×3

Step 3   Use the rule to find the number of dots in the next figure.

The next figure will have 3 more dots than the 4th figure.

It will have 5 rows of 3 dots. $5 \times 3 = 15$

Solution   **There will be 15 dots in the 5th figure.**

Coached Example

A number pattern has 6 terms. The first term is 55. The rule of the pattern is "subtract 9." What are the six terms in the number pattern?

Identify the information given in the problem.

The pattern has _____ terms.

The pattern starts with _____.

The rule is _____.

Use the rule to find the rest of the terms.

Subtract _____ from 55.

_____ − _____ = _____　　　←　second term

_____ − _____ = _____　　　←　third term

_____ − _____ = _____　　　←　fourth term

_____ − _____ = _____　　　←　fifth term

_____ − _____ = _____　　　←　sixth term

The six terms in the number pattern are _____, _____, _____, _____, _____, and _____.

Lesson Practice

Choose the correct answer.

1. What is the rule of this pattern?

   22 28 34 40 46 52

   A. add 4

   B. multiply by 4

   C. add 6

   D. multiply by 6

2. Which follows the rule "subtract 7"?

   A. 72   65   59   53   46

   B. 81   74   67   60   53

   C. 76   69   64   58   52

   D. 71   78   85   92   99

3. What is the next number in this pattern?

   6   14   22   30   38   ?

   A. 42

   B. 46

   C. 48

   D. 51

4. What is the missing number in this pattern?

   95   89   83   77   ?   65

   A. 79

   B. 73

   C. 71

   D. 68

5. A number pattern has 5 terms. The pattern starts with 27. The rule of the pattern is "add 4." Which could be the pattern?

   A. 27   31   39   51   67

   B. 27   23   19   15   11

   C. 27   31   35   39   43

   D. 27   28   30   33   37

6. What is the next number in this pattern?

   1   3   9   27   81   ?

   A. 97

   B. 99

   C. 135

   D. 243

7. If the pattern continues, how many dots will be in the next figure?

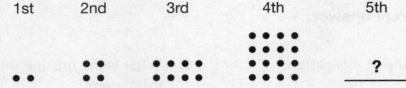

A. 32

B. 24

C. 20

D. 16

8. Which is the 14th figure in this pattern?

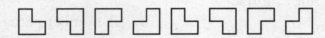

A.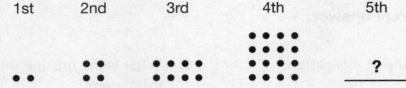

B.

C.

D.

9. Nelson wrote two number patterns. Each pattern has six terms. The first term in both patterns is 115.

A. Nelson used the rule "add 13" in the first pattern. What are the six terms in this pattern?

_____

B. Nelson used the rule "subtract 4" in the second pattern. What are the six terms in this pattern?

_____

Domain 2: Cumulative Assessment for Lessons 11–17

1. Which is **not** a factor pair of 56?

   A. {1, 56}

   B. {2, 28}

   C. {7, 8}

   D. {4, 13}

2. A TV show had 16,982 viewers last Tuesday. This Tuesday, the show had 3,472 more viewers than last Tuesday. How many viewers did the show have this Tuesday?

   A. 51,702

   B. 20,454

   C. 19,354

   D. 13,510

3. Which is a multiple of 9?

   A. 19

   B. 28

   C. 64

   D. 72

4. Marvin spent $827 on a computer. To the nearest hundred, what is the cost of the computer?

   A. $800   C. $830

   B. $820   D. $900

5. The tallest building in the world is 2,716 feet high. To the nearest thousand, what is the height of the tallest building?

   A. 2,700 feet

   B. 2,720 feet

   C. 2,800 feet

   D. 3,000 feet

6. The table shows the distances traveled by 3 students during the summer months.

   Distances Traveled

Student	Distance Traveled (in miles)
Hank	2,683
Nina	5,893
Ruby	453

   Which is the best estimate of how many more miles Nina traveled than Ruby?

   A. 2,000 miles

   B. 5,000 miles

   C. 5,500 miles

   D. 6,000 miles

7. Franklin typed for a total of 88 minutes. He can type 38 words in one minute. To find how many words he typed in all, which rounded numbers can you use to check the answer?

A. 80 × 30

B. 80 × 40

C. 90 × 30

D. 90 × 40

8. Cheryl's car traveled 226 miles on 8 gallons of gas. She traveled about the same number of miles with each gallon. Which is the best estimate for the number of miles she traveled with each gallon of gas?

A. 20 miles

B. 30 miles

C. 40 miles

D. 60 miles

9. A number pattern has 5 terms. The first term is 13. The rule of the pattern is "add 9." What are the 5 terms of the pattern?

_____

10. The table shows the number of species of some animal groups.

Number of Species

Animal Group	Number of Species
Birds	9,956
Mammals	5,416
Amphibians	6,199
Reptiles	8,240

A. How many species of mammals and amphibians are there?

_____

B. How many more species of birds than reptiles are there?

_____

_____

Domain 3 | Fractions

Domain 3: Diagnostic Assessment for Lessons 18–27

Lesson 18 Equivalent Fractions
4.NF.1

Lesson 19 Mixed Numbers and
Improper Fractions
4.NF.1

Lesson 20 Compare Fractions
4.NF.2

Lesson 21 Add Fractions
4.NF.3.a, 4.NF.3.b, 4.NF.3.d

Lesson 22 Subtract Fractions
4.NF.3.a, 4.NF.3.d

Lesson 23 Add and Subtract Mixed
Numbers
4.NF.3.b, 4.NF.3.c

Lesson 24 Multiply Fractions with
Whole Numbers
4.NF.4.a, 4.NF.4.b, 4.NF.4.c

Lesson 25 Decimals
4.NF.6

Lesson 26 Relate Decimals to
Fractions
4.NF.5, 4.NF.6

Lesson 27 Compare and Order
Decimals
4.NF.7

Domain 3: Cumulative Assessment for Lessons 18–27

Domain 3: Diagnostic Assessment for Lessons 18–27

1. Vienna drew this rectangle.

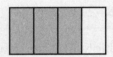

Which fraction is equivalent to the shaded part of Vienna's rectangle?

A. $\frac{3}{6}$ **C.** $\frac{6}{8}$

B. $\frac{7}{10}$ **D.** $\frac{8}{12}$

2. Karim walked $\frac{3}{10}$ mile to Fred's apartment. Then he walked $\frac{4}{10}$ mile to the park. How far did Karim walk in all?

A. $\frac{1}{10}$ mile

B. $\frac{7}{10}$ mile

C. $\frac{12}{10}$ mile

D. $\frac{7}{20}$ mile

3. Anthony has a piece of paper that is $\frac{11}{12}$ inch long. He cuts $\frac{1}{4}$ inch off the length of it. How long is the paper now?

A. $\frac{1}{2}$ inch **C.** $\frac{8}{12}$ inch

B. $\frac{5}{8}$ inch **D.** $\frac{9}{12}$ inch

4. Multiply.

$$3 \times \frac{3}{5} = \boxed{}$$

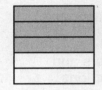

A. $\frac{6}{5}$

B. $\frac{9}{5}$

C. $\frac{5}{9}$

D. $\frac{9}{15}$

5. Which symbol makes this sentence true?

$$\frac{2}{8} \bigcirc \frac{1}{3}$$

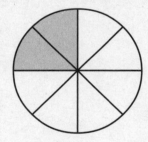 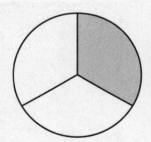

A. >

B. <

C. =

D. +

6. Which decimal represents the shaded parts of the grids?

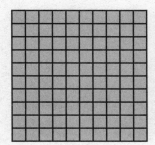 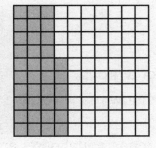

A. 0.36

C. 1.36

B. 1.30

D. 1.40

7. It rained 0.7 of the days of Omar's vacation. What fraction of the days did it rain?

A. $\frac{3}{100}$

C. $\frac{3}{10}$

B. $\frac{7}{100}$

D. $\frac{7}{10}$

8. Which lists the decimals from least to greatest?

A. 0.50   0.58   0.80

B. 0.50   0.80   0.58

C. 0.58   0.80   0.50

D. 0.80   0.58   0.50

9. What mixed number goes in the box to make the sentence true?

$$\boxed{} - 3\frac{4}{10} = 5\frac{3}{10}$$

---

10. Shelly bought $\frac{6}{8}$ pound of potato salad and $\frac{4}{8}$ pound of macaroni salad.

A. How many pounds of salad did she buy in all? Show your work.

B. Shelly also bought $2\frac{2}{8}$ pound of tuna salad. How many more pounds of tuna salad did she buy than macaroni salad? Show your work.

Equivalent Fractions

Common Core State Standard:
4.NF.1

Getting the Idea

A **fraction** names parts of a whole or part of a group.

The **denominator**, the bottom number, tells how many equal parts in the whole or group.

The **numerator**, the top number, tells how many equal parts are being considered.

The diagram below shows 7 equal parts in the whole. Each part is $\frac{1}{7}$ of the whole. $\frac{4}{7}$ of the figure is shaded. You read the fraction as *four sevenths*.

$\frac{1}{7}$	$\frac{1}{7}$	$\frac{1}{7}$	$\frac{1}{7}$	$\frac{1}{7}$	$\frac{1}{7}$	$\frac{1}{7}$

Example 1

What fraction of the circle is shaded?

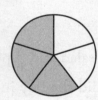

Strategy   **Find the denominator and the numerator.**

Step 1   Count the number of equal parts in the circle.

There are 5 equal parts. The denominator is 5.

Step 2   Count the number of shaded parts in the circle.

There are 3 shaded parts. The numerator is 3.

Step 3   Write the fraction.

$$\frac{\text{numerator}}{\text{denominator}} = \frac{3}{5}$$

Solution   **Three-fifths, or $\frac{3}{5}$, of the circle is shaded.**

Equivalent fractions are fractions that name the same value, but have different numerators and denominators. The models below show that $\frac{1}{2}$ and $\frac{2}{4}$ are equivalent fractions.

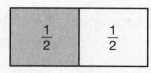

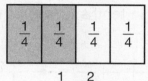

$$\frac{1}{2} = \frac{2}{4}$$

Example 2

What fraction with 12 as a denominator is equivalent to $\frac{2}{3}$?

$$\frac{2}{3} = \frac{\boxed{8}}{12}$$

Strategy   **Use fraction strips.**

Step 1   Use $\frac{1}{3}$ fraction strips.

Show $\frac{2}{3}$.

$\frac{1}{3}$	$\frac{1}{3}$	

Step 2   Use $\frac{1}{12}$ fraction strips.

Use as many strips as needed to make the same length as $\frac{2}{3}$.

$\frac{1}{3}$	$\frac{1}{3}$	

$\frac{1}{12}$	$\frac{1}{12}$	$\frac{1}{12}$	$\frac{1}{12}$	$\frac{1}{12}$	$\frac{1}{12}$	$\frac{1}{12}$	$\frac{1}{12}$	

Step 3   Count the number of $\frac{1}{12}$ strips.

There are 8 strips of $\frac{1}{12}$.

So, $\frac{8}{12}$ is equal to $\frac{2}{3}$.

Solution $\frac{2}{3} = \frac{8}{12}$

You can use number lines to find equivalent fractions.

Example 3

What fraction with 5 as a denominator is equivalent to $\frac{6}{10}$?

Strategy   **Use number lines.**

Step 1   Make a number line in tenths and another one in fifths.
Find $\frac{6}{10}$.

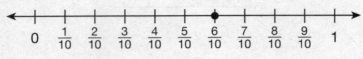

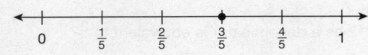

Step 2   Find the fraction on the number line in fifths that lines up with $\frac{6}{10}$.

$\frac{3}{5}$ lines up with $\frac{6}{10}$.

Solution $\frac{3}{5}$ is equivalent to $\frac{6}{10}$.

Another way to find equivalent fractions is to multiply the numerator and the denominator by the same number. For example:

$$\frac{3}{5} = \frac{3 \times 2}{5 \times 2} = \frac{6}{10}$$

Example 4

What is an equivalent fraction of $\frac{3}{4}$?

Strategy   **Multiply the numerator and denominator by the same number.**

Step 1   Multiply both the numerator and denominator by 2.

$$\frac{3}{4} \times \frac{2}{2} = \frac{3 \times 2}{4 \times 2} = \frac{6}{8}$$

Step 2   Use models to check.

The models for $\frac{3}{4}$ and $\frac{6}{8}$ have the same size.

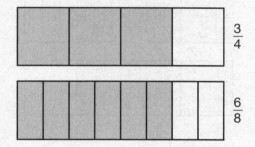

$\frac{3}{4}$

$\frac{6}{8}$

Solution $\frac{6}{8}$ **is an equivalent fraction of** $\frac{3}{4}$.

A fraction is in **simplest form** if its numerator and denominator have only 1 as a common factor. A fraction and its simplest form are equivalent fractions.

You can simplify a fraction by dividing the numerator and denominator by the greatest common factor. The **greatest common factor (GCF)** is the greatest factor that is common to two or more numbers.

For example, simplify $\frac{6}{8}$.

2 is the greatest common factor (GCF) of 6 and 8.

So, divide the numerator and the denominator by 2.

$$\frac{6}{8} = \frac{6 \div 2}{8 \div 2} = \frac{3}{4}$$

$\frac{3}{4}$ is the simplest form of $\frac{6}{8}$.

Example 5

Simplify the fraction $\frac{9}{12}$.

Strategy   **Divide the numerator and denominator by the GCF.**

Step 1   Find the greatest common factor of 9 and 12.

           The GCF is 3.

Step 2   Divide both the numerator and denominator by 3.

$$\frac{9}{12} = \frac{9 \div 3}{12 \div 3} = \frac{3}{4}$$

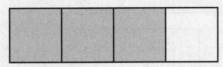

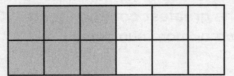

Solution $\frac{9}{12}$ in simplest form is $\frac{3}{4}$.

Coached Example

What fraction with 12 as a denominator is equivalent to $\frac{3}{6}$?

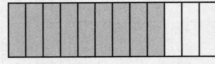

What is the denominator of $\frac{3}{6}$? _____

What is the denominator of the equivalent fraction? _____

By what number can you multiply 6 to get 12? _____

To find the equivalent fraction, multiply the numerator and denominator by

_____.

$$\frac{3 \times \rule{1.5cm}{0.4pt}}{6 \times \rule{1.5cm}{0.4pt}} = \rule{1.5cm}{0.4pt}$$

_____ is a fraction with 12 as a denominator that is equivalent to $\frac{3}{6}$.

Lesson Practice

Choose the correct answer.

1. Which shows $\frac{3}{4}$ of the figure shaded?

   A.

   B.

   C.

   D.

2. The fraction strips show $\frac{10}{12}$.

   Which fraction is equivalent to $\frac{10}{12}$?

   A. $\frac{1}{6}$

   B. $\frac{1}{3}$

   C. $\frac{2}{3}$

   D. $\frac{5}{6}$

3. The model is shaded to represent a fraction.

   Which model shows an equivalent fraction?

   A.

   B.

   C.

   D.

4. What fraction with 9 as a denominator is equivalent to $\frac{2}{3}$?

   A. $\frac{1}{9}$ **C.** $\frac{4}{9}$

   B. $\frac{2}{9}$ **D.** $\frac{6}{9}$

5. What fraction with 3 as a denominator is equivalent to $\frac{8}{12}$?

A. $\frac{1}{3}$

B. $\frac{2}{3}$

C. $\frac{3}{3}$

D. $\frac{4}{3}$

6. Which fraction is **not** equivalent to $\frac{1}{2}$?

A. $\frac{5}{10}$

B. $\frac{4}{8}$

C. $\frac{3}{12}$

D. $\frac{2}{4}$

7. Which fraction is written in simplest form?

A. $\frac{1}{5}$

B. $\frac{6}{8}$

C. $\frac{3}{9}$

D. $\frac{4}{12}$

8. Which is the simplest form of $\frac{2}{8}$?

A. $\frac{1}{4}$

B. $\frac{1}{8}$

C. $\frac{2}{4}$

D. $\frac{4}{16}$

9. Jonah made the figure below.

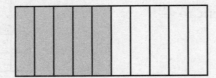

A. What fraction of the figure is shaded?

_____

B. Write two equivalent fractions to the fraction in Part A.

_____

_____

Common Core State Standard:
4.NF.1

Mixed Numbers and Improper Fractions

Getting the Idea

An **improper fraction** is a fraction that has a numerator equal to or greater than the denominator. Examples of improper fractions are $\frac{3}{2}$, $\frac{5}{5}$, and $\frac{7}{4}$.

Example 1

What improper fraction does this model represent?

| $\frac{1}{4}$ | $\frac{1}{4}$ | $\frac{1}{4}$ | $\frac{1}{4}$ | | $\frac{1}{4}$ | $\frac{1}{4}$ | $\frac{1}{4}$ | $\frac{1}{4}$ | | $\frac{1}{4}$ | $\frac{1}{4}$ | $\frac{1}{4}$ | $\frac{1}{4}$ |

Strategy   **Find the denominator. Then find the numerator.**

Step 1   What kind of fraction bars are used in the model?

Each bar is in fourths. The denominator will be 4.

Step 2   Count the number of shaded parts.

There are 11 shaded parts. The numerator will be 11.

Step 3   Write the numerator over the denominator.

$\frac{11}{4}$

Solution   The model represents the improper fraction $\frac{11}{4}$.

A **mixed number** has a whole number part and a fraction part. You use a mixed number when you say your age is $9\frac{1}{2}$ or your shoe size is $5\frac{1}{2}$.

Example 2

What mixed number does this model represent?

| $\frac{1}{5}$ | $\frac{1}{5}$ | $\frac{1}{5}$ | $\frac{1}{5}$ | $\frac{1}{5}$ |

| $\frac{1}{5}$ | $\frac{1}{5}$ | $\frac{1}{5}$ | $\frac{1}{5}$ | $\frac{1}{5}$ |

| $\frac{1}{5}$ | $\frac{1}{5}$ | $\frac{1}{5}$ | $\frac{1}{5}$ | $\frac{1}{5}$ |

Strategy   **Find the whole number part. Then find the fraction part.**

Step 1   Find the whole number part.

There are 2 fraction bars completely shaded.

Remember that $\frac{5}{5} = 1$.

The whole number part is 2.

Step 2   Find the fraction part.

The third fraction bar has 3 shaded parts out of 5 parts.

The fraction part is $\frac{3}{5}$.

Solution   **The model represents the mixed number $2\frac{3}{5}$.**

Example 3

Where is point *C* located on the number line? Write the answer as an improper fraction and as a mixed number.

Strategy   **Decide what each tick mark represents. Then count.**

Step 1   What does each tick mark represent?

Each tick mark shows $\frac{1}{4}$.

Step 2   To find the improper fraction, count the number of tick marks past 0.

Point *C* is 10 tick marks to the right of 0.

The improper fraction is $\frac{10}{4}$.

Step 3   To find the mixed number, count the number of tick marks past the whole number.

Point *C* is two marks to the right of 2.

The mixed number $2\frac{2}{4}$.

Solution   **Point *C* is located at $\frac{10}{4}$ or $2\frac{2}{4}$ on the number line.**

You can change an improper fraction to a mixed number and a mixed number to an improper fraction. To change a mixed number to an improper fraction, multiply the whole number part by the denominator. Then add the numerator to that product. The denominator stays the same.

Example 4

Jamie brought a batch of pies to a potluck dinner. Each pie was cut into 6 equal pieces. After dinner, there were $3\frac{5}{6}$ pies left over. What is $3\frac{5}{6}$ as an improper fraction?

Strategy　　**Use multiplication and addition.**

Step 1　　Multiply the whole number part by the denominator of that fraction.

$$3 \times 6 = 18$$

Step 2　　Add that product to the numerator of the fraction part.

$$18 + 5 = 23$$

Step 3　　The denominator stays the same. Write the improper fraction.

$$\frac{23}{6}$$

Solution　　$3\frac{5}{6} = \frac{23}{6}$

To change an improper fraction to a mixed number, divide the numerator by the denominator. The quotient is the whole number; the remainder is the numerator of the fraction. The denominator stays the same.

Example 5

Flora made placemats using $\frac{13}{3}$ yards of fabric. What is $\frac{13}{3}$ as a mixed number?

Strategy　　**Divide the numerator by the denominator.**

Step 1　　Divide the numerator by the denominator.

$$13 \div 3 = 4 \text{ R1}$$

Step 2　　The quotient, 4, is the whole number part.

The remainder, 1, is the numerator of the fraction part.

The denominator, 3, stays the same.

Write the fraction part: $\frac{1}{3}$

Solution　　$\frac{13}{3} = 4\frac{1}{3}$

Coached Example

Write an improper fraction and a mixed number to represent this model.

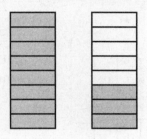

First, write an improper fraction.

Each figure is divided into _____ parts.

The denominator of the improper fraction is _____.

There are _____ shaded parts.

The improper fraction is _____.

Now write the mixed number.

How many figures are completely shaded? _____

The whole number part of the mixed number is _____.

The second figure has _____ parts in all and _____ shaded parts.

The fraction part of the mixed number is _____.

The mixed number is _____.

The model represents _____ or _____.

Lesson Practice

Choose the correct answer.

1. Look at the model below.

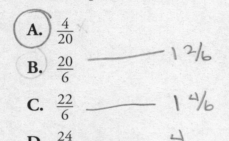

   What improper fraction does the model represent?

   A. $\frac{4}{20}$ ✗

   B. $\frac{20}{6}$ ——— $1\,2/6$

   C. $\frac{22}{6}$ ——— $1\,4/6$

   D. $\frac{24}{6}$ ——— 4

2. Which improper fraction is equal to $3\frac{1}{2}$?

   A. $\frac{6}{3}$

   B. $\frac{7}{3}$ $3\,1/2 = 7/2$

   C. $\frac{6}{2}$

   D. $\frac{7}{2}$

3. Hayley ordered several pizzas for a party. The picture below shows the amount of pizza that is left after the party.

   How much pizza is left after the party?

   A. $1\frac{1}{5}$

   B. $1\frac{5}{8}$

   C. $8\frac{3}{8}$

   D. $8\frac{5}{8}$

4. Which improper fraction can be written as a whole number?

   A. $\frac{10}{6}$

   B. $\frac{12}{6}$

   C. $\frac{14}{6}$

   D. $\frac{16}{6}$

5. Look at the number line below.

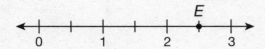

What mixed number is shown by point *E* on the number line?

A. $1\frac{1}{2}$

B. $2\frac{1}{4}$

C. $2\frac{1}{2}$

D. $2\frac{3}{4}$

6. Look at the number line below.

What mixed number is shown by point *F* on the number line?

A. $\frac{9}{4}$

B. $\frac{11}{4}$

C. $\frac{12}{4}$

D. $\frac{13}{4}$

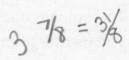

$3\,1/4 = 13/4$

7. Look at the model below.

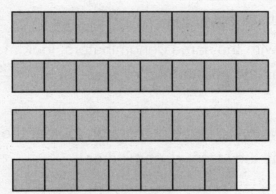

$3\ 7/8 = 3\frac{1}{8}$

A. Write an improper fraction to represent the model.

 31/8

B. Write a mixed number to represent the model.

3 7/8

Common Core State Standard:
4.NF.2

Compare Fractions

Getting the Idea

Remember to refer to the same whole, when comparing fractions.

For example,

$\frac{1}{2}$ of a square is equal to $\frac{1}{2}$ of the same size square.

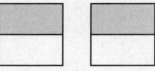

$\frac{1}{2}$ of a square is not equal to $\frac{1}{2}$ of a circle.

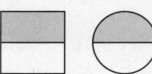

Use these symbols to compare fractions.

   The symbol $>$ means *is greater than*.

   The symbol $<$ means *is less than*.

   The symbol $=$ means *is equal to*.

To compare fractions with the same denominators, look at the numerators.

The greater fraction has the greater numerator.

For example, $\frac{2}{3} > \frac{1}{3}$.

To compare fractions with the same numerator, look at the denominators.

The greater fraction has the lesser denominator.

For example, $\frac{2}{3} > \frac{2}{5}$.

Example 1

Which symbol makes the sentence true? Write $>$, $<$, or $=$.

   $\frac{1}{8}$ $<$ $\frac{1}{4}$

Strategy   **Look at the denominators. Use fraction strips to check.**

   Step 1   Both fractions have the same numerator, 1.

   Compare the denominators. $8 > 4$

   Since 8 is the greater denominator, $\frac{1}{8}$ is the lesser fraction.

Step 2   Use fraction strips to check.

The $\frac{1}{8}$ fraction strip is shorter than $\frac{1}{4}$ fraction strip.

$\frac{1}{8}$							

$\frac{1}{4}$			

Step 3   Use the correct symbol.

< means is less than.

Solution $\frac{1}{8}$ $\textcircled{<}$ $\frac{1}{4}$

You can use number lines to help, when comparing fractions.
The fraction farther to the right is the greater fraction.

Example 2

Which symbol makes the sentence true? Write >, <, or =.

$\frac{2}{3}$ $\bigcirc$ $\frac{7}{12}$

Strategy   **Use number lines.**

Step 1   Draw two number lines from 0 to 1.

Make one number line in thirds. Find $\frac{2}{3}$.

Make another number line in twelfths. Find $\frac{7}{12}$.

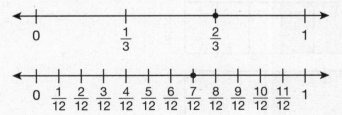

Step 2   Compare the fractions.

$\frac{2}{3}$ is farther to the right than $\frac{7}{12}$.

So, $\frac{2}{3}$ is the greater fraction.

Solution $\frac{2}{3}$ $\textcircled{>}$ $\frac{7}{12}$

For Example 2, you can also use a common denominator to compare $\frac{2}{3}$ and $\frac{7}{12}$.
A **common denominator** is a common multiple of the denominators.

The least common multiple (LCM) of 3 and 12 is 12.

Change $\frac{2}{3}$ to an equivalent fraction with 12 as the denominator.

$$\frac{2}{3} = \frac{2 \times 4}{3 \times 4} = \frac{8}{12}$$

Now compare the numerators.

Since 8 is greater, $\frac{8}{12} > \frac{7}{12}$. So, $\frac{2}{3} > \frac{7}{12}$.

Example 3

Shari and Billy each bought the same candy bar. Shari ate $\frac{2}{5}$ of her candy bar.

Billy ate $\frac{1}{2}$ of his candy bar. Who ate more of their candy bar?

Strategy   **Use a common denominator.**

Step 1   Find a common denominator for $\frac{2}{5}$ and $\frac{1}{2}$.
The LCM of 5 and 2 is 10.

Step 2   Change $\frac{2}{5}$ and $\frac{1}{2}$ to equivalent fractions with 10 as the denominator.

$$\frac{2}{5} = \frac{2 \times 2}{5 \times 2} = \frac{4}{10} \qquad \frac{1}{2} = \frac{1 \times 5}{2 \times 5} = \frac{5}{10}$$

Step 3   Compare the numerators.

Since 4 is the lesser numerator, $\frac{4}{10} < \frac{5}{10}$. So $\frac{2}{5} < \frac{1}{2}$.

Step 4   Use models to check.

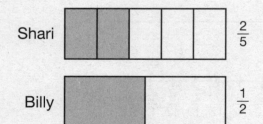

Solution   **Billy ate more of his candy bar.**

You can also use common numerators to compare fractions.

A common numerator is a common multiple of the numerators.

Example 4

Lilly needs $\frac{3}{8}$ cup of milk and $\frac{1}{3}$ cup of heavy cream to make pancakes.

Does she need more milk or more heavy cream to make the pancakes?

Strategy   **Use a common numerator.**

| Step 1 | Find a common numerator for $\frac{3}{8}$ and $\frac{1}{3}$. |

The LCM of 3 and 1 is 3.

| Step 2 | Change $\frac{1}{3}$ to an equivalent fraction with 3 as the numerator. |

$$\frac{1}{3} = \frac{1 \times 3}{3 \times 3} = \frac{3}{9}$$

$\frac{3}{8}$ already has a 3 as the numerator.

| Step 3 | Compare the denominators. |

The fraction with the lesser denominator is the greater fraction.

Since 8 is the lesser denominator, $\frac{3}{8} > \frac{3}{9}$.

So $\frac{3}{8} > \frac{1}{3}$.

| Step 4 | Use models to check. |

| $\frac{1}{8}$ | $\frac{1}{8}$ | $\frac{1}{8}$ | | | | | | $\frac{3}{8}$ |

| $\frac{1}{3}$ | | | $\frac{1}{3}$ |

Solution   **Lilly needs more milk to make the pancakes.**

Fractions can be compared using **benchmarks**. A benchmark is a common number that can be compared to another number. Three benchmarks are 0, $\frac{1}{2}$, and 1. You can think of a number line when using a benchmark.

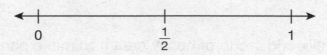

Example 5

What symbol makes this sentence true? Write >, <, or =.

$$\frac{4}{10} \bigcirc \frac{5}{6}$$

Strategy Use $\frac{1}{2}$ as a benchmark.

Step 1 Find $\frac{4}{10}$ on a number line.

Compare $\frac{4}{10}$ to $\frac{1}{2}$.

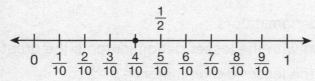

$\frac{4}{10}$ is less than $\frac{1}{2}$.

Step 2 Find $\frac{5}{6}$ on a number line.

Compare $\frac{5}{6}$ to $\frac{1}{2}$.

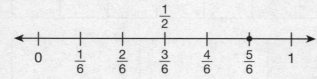

$\frac{5}{6}$ is greater than $\frac{1}{2}$.

Step 3 Compare $\frac{5}{6}$ to $\frac{4}{10}$.

Since $\frac{4}{10}$ is less than $\frac{1}{2}$ and $\frac{5}{6}$ is greater than $\frac{1}{2}$,

$\frac{4}{10}$ is less than $\frac{5}{6}$.

Solution $\frac{4}{10} \bigcirc< \frac{5}{6}$

Coached Example

What symbol makes this sentence true?
Use benchmarks 0, $\frac{1}{2}$, and 1 to compare.

Write >, <, or =.

$\frac{4}{5}$ ◯ $\frac{1}{6}$

Find $\frac{4}{5}$ on a number line.

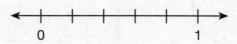

Compare to 0, $\frac{1}{2}$, and 1 on a number line.

$\frac{4}{5}$ is closest to the benchmark _____.

Find $\frac{1}{6}$ on a number line.

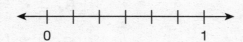

Compare to 0, $\frac{1}{2}$, and 1 on the number line.

$\frac{1}{6}$ is closest to the benchmark _____.

Since $\frac{4}{5}$ is closest to the benchmark _____, and $\frac{1}{6}$ is closest to the benchmark

_____, $\frac{4}{5}$ is _____ than $\frac{1}{6}$.

$\frac{4}{5}$ ◯ $\frac{1}{6}$

Lesson Practice

Choose the correct answer.

1. Which symbol makes this sentence true?

   $\frac{2}{6} \bigcirc \frac{1}{5}$

   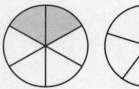

   A. >

   B. <

   C. =

   D. +

2. Which symbol makes this sentence true?

   $\frac{3}{4} \bigcirc \frac{3}{8}$

   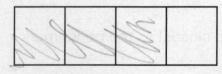

   A. >

   B. <

   C. =

   D. +

3. Which sentence is true?

   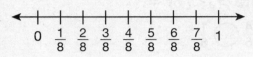

   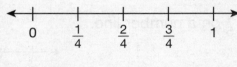

   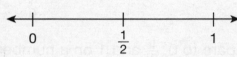

   A. $\frac{1}{8} > \frac{1}{4}$

   B. $\frac{1}{2} < \frac{2}{4}$

   C. $\frac{3}{8} > \frac{1}{2}$

   D. $\frac{5}{8} < \frac{3}{4}$

4. Which symbol makes this sentence true?

   $\frac{7}{12} \bigcirc \frac{3}{4}$

   A. >   C. =

   B. <   D. +

5. Which sentence is **not** true?

   A. $\frac{1}{5} > \frac{1}{8}$ C. $\frac{1}{7} < \frac{1}{2}$

   B. $\frac{1}{3} > \frac{1}{6}$ D. $\frac{1}{4} < \frac{1}{9}$

6. Which fraction makes this sentence true?

$$\frac{3}{4} < \underline{\hspace{2cm}}$$

A. $\frac{1}{2}$

B. $\frac{7}{8}$

C. $\frac{5}{10}$

D. $\frac{7}{12}$

7. Which fraction is closest to 1?

A. $\frac{2}{3}$ C. $\frac{7}{8}$

B. $\frac{3}{4}$ D. $\frac{7}{10}$

8. Which symbol makes this sentence true?

$$\frac{3}{10} \bigcirc \frac{25}{100}$$

A. >   C. =

B. <   D. +

---

9. Marjorie used $\frac{2}{5}$ cup of flour, $\frac{1}{4}$ cup of baking soda, and $\frac{1}{3}$ cup of sugar for a recipe.

A. Did Marjorie use more baking soda or more sugar? Show your work.

sugar

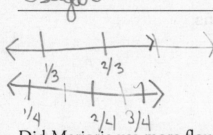

B. Did Marjorie use more flour or more sugar for the recipe? Show your work.

flour

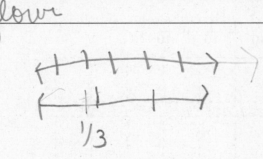

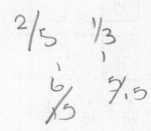

Add Fractions

Common Core State Standards:
4.NF.3.a, 4.NF.3.b, 4.NF.3.d

Getting the Idea

You can show a fraction as an equation.

$$\frac{3}{4} = \frac{1}{4} + \frac{1}{4} + \frac{1}{4}$$

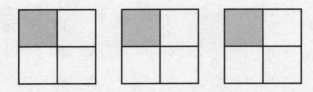

Example 1

Show the fraction $\frac{3}{10}$ as an equation.

Make a model to show the equation.

Strategy　　**Break the fraction into unit fractions.**

Step 1　Show the fraction as unit fractions.

$$\frac{3}{10} = \frac{1}{10} + \frac{1}{10} + \frac{1}{10}$$

Step 2　Make a model.

Draw a rectangle. Show the rectangle in tenths.

$$\frac{3}{10} = \frac{1}{10} + \frac{1}{10} + \frac{1}{10}$$

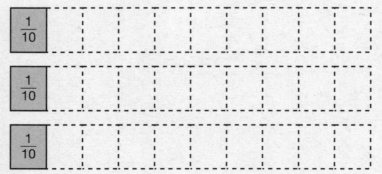

Solution　$\frac{3}{10} = \frac{1}{10} + \frac{1}{10} + \frac{1}{10}$

You can add fractions that have the same denominators, which are called **like denominators**. If fractions have like denominators, add the numerators. The denominator stays the same. If the sum is an improper fraction, rewrite it as a mixed number.

Example 2

Add.

$$\frac{4}{6} + \frac{1}{6} = \boxed{}$$

Strategy   **Add like fractions.**

Step 1   Do the fractions have a like denominator?

Yes, both fractions have a denominator of 6.

Step 2   Add the numerators.

4 + 1 = 5

Step 3   The denominator stays the same.

$$\frac{4}{6} + \frac{1}{6} = \frac{5}{6}$$

Step 4   Use an area model to check the sum.

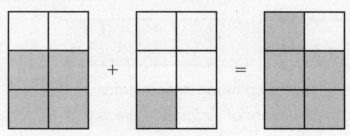

Solution $\frac{4}{6} + \frac{1}{6} = \frac{5}{6}$

Example 3

Allison bought a pizza divided into 10 equal slices. Allison's friends ate $\frac{7}{10}$ of the pizza and her brother ate $\frac{2}{10}$ of the pizza. What fraction of the pizza was eaten in all?

Strategy   **Use fraction strips to model the problem.**

Step 1   Write an addition sentence for the problem.

The words "in all" tell you to add.

Allison's friends ate $\frac{7}{10}$ pizza.

Allison's brother ate $\frac{2}{10}$ pizza.

Let p represent the total pizza eaten.

$\frac{7}{10} + \frac{2}{10} = p$

Step 2   Show $\frac{7}{10}$ with fraction pieces.

| $\frac{1}{10}$ | $\frac{1}{10}$ | $\frac{1}{10}$ | $\frac{1}{10}$ | $\frac{1}{10}$ | $\frac{1}{10}$ | $\frac{1}{10}$ | | | |

Step 3   Add on $\frac{2}{10}$ with fraction pieces.

| $\frac{1}{10}$ | $\frac{1}{10}$ | $\frac{1}{10}$ | $\frac{1}{10}$ | $\frac{1}{10}$ | $\frac{1}{10}$ | $\frac{1}{10}$ | $\frac{1}{10}$ | $\frac{1}{10}$ | |

Step 4   The fractions have a common denominator.

Count the numbers of $\frac{1}{10}$ fraction pieces.

There are nine $\frac{1}{10}$ fraction pieces in all.

| $\frac{1}{10}$ | $\frac{1}{10}$ | $\frac{1}{10}$ | $\frac{1}{10}$ | $\frac{1}{10}$ | $\frac{1}{10}$ | $\frac{1}{10}$ | $\frac{1}{10}$ | $\frac{1}{10}$ | |

$\frac{1}{10} , \frac{2}{10} , \frac{3}{10} , \frac{4}{10} , \frac{5}{10} , \frac{6}{10} , \frac{7}{10} , \frac{8}{10} , \frac{9}{10}$

Step 5   Write the sum.

$\frac{7}{10} + \frac{2}{10} = \frac{9}{10}$

Solution   **Allison's friends and brother ate $\frac{9}{10}$ of the pizza in all.**

Example 4

Find the sum.

$$\frac{5}{8} + \frac{7}{8} = \boxed{}$$

Strategy	**Add the numerators and use the common denominator.**

Step 1   Add the numerators. The denominator stays the same.

$$\frac{5}{8} + \frac{7}{8} = \frac{5+7}{8} = \frac{12}{8}$$

Step 2   Change the improper fraction to a mixed number.

Divide the numerator by the denominator.

$$12 \div 8 = 1\ R4$$

Step 3   Write the mixed number.

The quotient is the whole number.

The remainder is the numerator.

The denominator stays the same.

So, $\frac{12}{8} = 1\frac{4}{8}$.

Solution $\frac{5}{8} + \frac{7}{8} = 1\frac{4}{8}$

Coached Example

Sam walked $\frac{1}{12}$ mile to Toni's house. Then Toni and Sam walked $\frac{4}{12}$ mile to school. How far did Sam walk in all?

Write the addition sentence for the problem.

The words "in all" tell you to _____.

Sam walked _____ mile to Toni's house.

Then Sam walked _____ mile to school.

Let m represent the total miles Sam walked.

_____ + _____ = _____

Find the sum.

Do the fractions have a like denominator?

_____, both fractions have a denominator of _____.

Add the numerators.

1 + 4 = _____

The denominator stays the same.

$\frac{1}{12} + \frac{4}{12} = \frac{\square}{\square}$

Sam walked _____ mile in all.

Lesson Practice

Choose the correct answer.

1. Add.

$$\frac{2}{5} + \frac{1}{5} = \square$$

$\frac{1}{5}$	$\frac{1}{5}$	$\frac{1}{5}$	$\frac{1}{5}$	$\frac{1}{5}$

A. $\frac{3}{10}$

B. $\frac{6}{7}$

C. $\frac{3}{5}$

D. $\frac{1}{5}$

2. Add.

$$\frac{4}{7} + \frac{3}{7} = \square$$

A. $\frac{1}{7}$

B. $\frac{7}{14}$

C. $\frac{3}{4}$

D. 1

3. Add.

$$\frac{3}{10} + \frac{5}{10} = \square$$

A. $\frac{2}{10}$

B. $\frac{6}{10}$

C. $\frac{8}{10}$

D. $\frac{10}{10}$

4. Natalie ate $\frac{3}{8}$ of an apple. Later, she ate another $\frac{3}{8}$ of the apple. What fraction of the apple did she eat in all?

A. $\frac{2}{8}$

B. $\frac{6}{8}$

C. $\frac{6}{16}$

D. $\frac{8}{8}$

5. Find the sum.

$$\frac{3}{9} + \frac{1}{9} = \square$$

A. $\frac{2}{9}$

B. $\frac{2}{12}$

C. $\frac{4}{9}$

D. $\frac{4}{18}$

6. Find the sum.

$$\frac{5}{10} + \frac{3}{10} = \square$$

A. $\frac{4}{12}$

B. $\frac{4}{10}$

C. $\frac{8}{10}$

D. $\frac{8}{20}$

7. Manny spent $\frac{2}{6}$ hour reading and $\frac{3}{6}$ hour studying for his math quiz. How long did Manny spend reading and studying for the quiz?

   A. $\frac{5}{6}$ hour

   B. $\frac{4}{9}$ hour

   C. $\frac{4}{6}$ hour

   D. $\frac{2}{3}$ hour

8. Jamal did $\frac{2}{8}$ of his chores before lunch. He did another $\frac{3}{8}$ of his chores after lunch. What fraction of his chores did he finish?

   A. $\frac{4}{8}$

   B. $\frac{5}{8}$

   C. $\frac{4}{12}$

   D. $\frac{5}{16}$

9. The drive from Collin's house to pick up Anderson took $\frac{4}{10}$ tank of gas. Another $\frac{3}{10}$ tank of gas was used to drive from Anderson's house to the water park.

   A. How much gas was used to drive from Collin's house to the water park? Show your work.

   7/10

   4/10 + 3/10 = 7/10

   B. At the water park, Collin bought $\frac{6}{8}$ pound of fudge. How much fudge was left after Collin ate $\frac{1}{8}$ pound? Show your work.

   5/8

   6/8 - 1/8 = 5/8

Common Core State Standards:
4.NF.3.a, 4.NF.3.d

Subtract Fractions

Getting the Idea

When you subtract fractions with like denominators, subtract only the numerators. The denominator stays the same.

Example 1

Subtract.

$\frac{7}{8} - \frac{5}{8} = \boxed{}$

Strategy   **Use fraction strips to model the problem.**

Step 1   Show $\frac{7}{8}$ with fraction pieces.

| $\frac{1}{8}$ | $\frac{1}{8}$ | $\frac{1}{8}$ | $\frac{1}{8}$ | $\frac{1}{8}$ | $\frac{1}{8}$ | $\frac{1}{8}$ |

Step 2   Subtract or cross out $\frac{5}{8}$.

$\frac{5}{8} = \frac{1}{8} + \frac{1}{8} + \frac{1}{8} + \frac{1}{8} + \frac{1}{8}$

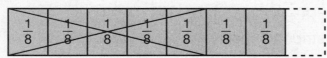

Step 3   Count the number of $\frac{1}{8}$ fraction pieces left.

There are two $\frac{1}{8}$ pieces left.

The numerator is 2.

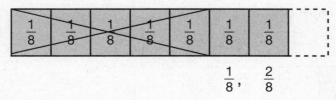

$\frac{1}{8}$, $\frac{2}{8}$

Step 4   The denominator is the same.

The denominator is 8.

Solution $\frac{7}{8} - \frac{5}{8} = \frac{2}{8}$

Sometimes the answer may not be in simplest form. Remember to simplify the fraction using the greatest common factor (GCF).

For example, for the fraction $\frac{3}{12}$, 3 is the greatest common factor of 3 and 12.

$$\frac{3}{12} = \frac{3 \div 3}{12 \div 3} = \frac{1}{4}$$

Example 2

Rachel used $\frac{5}{8}$ can of blue paint. She also used $\frac{3}{8}$ can of yellow paint.

How much more blue paint than yellow paint did Rachel use?

Strategy   **Subtract the numerators. Keep the same denominator.**

Step 1   Write the subtraction sentence for the problem.

The words "how much more" tell you to subtract.

She used $\frac{5}{8}$ can of blue paint.

She used $\frac{3}{8}$ can of yellow paint.

Let b represent how much more blue paint she used.

$$\frac{5}{8} - \frac{3}{8} = b$$

Step 2   The denominators are the same, so subtract the numerators.

$$\frac{5}{8} - \frac{3}{8} = \frac{5 - 3}{8} = \frac{2}{8}$$

Step 3   Use fraction strips to check.

Step 4   Simplify the fraction $\frac{2}{8}$.

2 is the GCF of 2 and 8.

$$\frac{2}{8} = \frac{2 \div 2}{8 \div 2} = \frac{1}{4}$$

Solution   Rachel used $\frac{2}{8}$ or $\frac{1}{4}$ can more blue paint than yellow paint.

Example 3

Tim had $\frac{10}{12}$ roll of green streamer for a bulletin board. He used $\frac{7}{12}$ roll for the border. What fraction of the roll of green streamer does Tim have left?

Strategy   **Subtract the numerators. Keep the same denominator.**

Step 1   Write the subtraction sentence for the problem.

The words "have left" tell you to subtract.

He had $\frac{10}{12}$ roll of streamer.

He used $\frac{7}{12}$ roll of streamer.

Let s represent how much is left over.

$$\frac{10}{12} - \frac{7}{12} = s$$

Step 2   The denominators are the same, so subtract the numerators.

$$\frac{10}{12} - \frac{7}{12} = \frac{10 - 7}{12} = \frac{3}{12}$$

Step 3   Simplify the fraction $\frac{3}{12}$.

3 is the GCF of 3 and 12.

$$\frac{3}{12} = \frac{3 \div 3}{12 \div 3} = \frac{1}{4}$$

Solution   **Tim has $\frac{3}{12}$ or $\frac{1}{4}$ roll of green streamer left.**

Remember, you can check the answer to a subtraction problem using addition.

$$\frac{3}{12} + \frac{7}{12} = \frac{10}{12}$$

The sum is $\frac{10}{12}$, which matches the minuend. So, the answer is correct.

Coached Example

Alexandra wants to jog $\frac{7}{10}$ mile. After jogging $\frac{3}{10}$ mile, her shoelaces become untied and she stops to retie them. How much more does Alexandra need to jog to finish her run?

Write the subtraction sentence for the problem.

The words "how much more" tell you to _____.

Alexandra wants to jog _____ mile.

She ties her shoelaces after _____ mile.

Let *m* represent how much more she needs to jog to finish her run.

_____ − _____ = _____

Find the difference.

Do the fractions have a like denominator?

_____, both fractions have a denominator of _____.

Subtract the numerators.

$7 - 3 =$ _____

The denominator stays the same.

$$\frac{7}{10} - \frac{3}{10} = \frac{\boxed{}}{\boxed{}}$$

Simplify the fraction. _____

Alexandra has _____ mile more to jog to finish her run.

Lesson Practice

Choose the correct answer.

1. Subtract.

$$\frac{7}{8} - \frac{2}{8} = \boxed{}$$

$\frac{1}{8}$	$\frac{1}{8}$	$\frac{1}{8}$	$\frac{1}{8}$	$\frac{1}{8}$	$\frac{1}{8}$	$\frac{1}{8}$	$\frac{1}{8}$

A. $\frac{4}{8}$ **C.** $\frac{0}{5}$

B. $\frac{5}{8}$ **D.** $\frac{1}{6}$

2. Subtract.

$$\frac{9}{10} - \frac{3}{10} = \boxed{}$$

A. $\frac{6}{20}$

B. $\frac{7}{10}$

C. $\frac{6}{10}$

D. $\frac{12}{20}$

3. Mary Ellen drew a rectangle with a length of $\frac{11}{12}$ inch. The height of the rectangle is $\frac{6}{12}$ inch shorter than the length. What is the height of the rectangle?

A. $\frac{0}{6}$ inch   **C.** $\frac{1}{7}$ inch

B. $\frac{5}{12}$ inch   **D.** $\frac{16}{12}$ inch

4. There was $\frac{3}{4}$ of a pad of paper on the counter. Greta used $\frac{1}{4}$ of the pad to write thank-you letters. What fraction of the pad of paper is left?

A. $\frac{1}{4}$

B. $\frac{1}{2}$

C. $\frac{3}{4}$

D. 1

5. Find the difference.

$$\frac{1}{2} - \frac{1}{2} = \boxed{}$$

A. 0

B. $\frac{1}{4}$

C. $\frac{1}{2}$

D. $\frac{2}{4}$

6. Find the difference.

$$\frac{65}{100} - \frac{34}{100} = \boxed{}$$

A. $\frac{21}{100}$

B. $\frac{31}{100}$

C. $\frac{41}{100}$

D. $\frac{99}{100}$

7. Jasmine ordered $\frac{3}{6}$ pound of American cheese and $\frac{5}{6}$ pound of cheddar cheese. How much more cheddar cheese than American cheese did Jasmine order?

   A. $\frac{1}{3}$ pound

   B. $\frac{1}{2}$ pound

   C. $\frac{3}{4}$ pound

   D. 1 pound

   $5/6 - 3/6 = 2/6$

   $2/6 = 1/3$

8. Kyle planted vegetables in $\frac{3}{8}$ of the space in his garden. He planted flowers in $\frac{2}{8}$ of the space in his garden. How much more of the garden space did Kyle plant with vegetables than with flowers?

   A. $\frac{0}{8}$ C. $\frac{1}{4}$

   B. $\frac{1}{8}$ D. $\frac{3}{8}$

9. The table shows the amount of fruit that Keisha put in a fruit salad.

Keisha's Fruit Salad

Fruit	Amount (in pounds)
Grapes	$\frac{3}{10}$
Blueberries	$\frac{2}{10}$
Watermelon	$\frac{8}{10}$

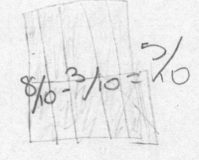

$8/10 - 3/10 = 5/10$

A. How many more pounds of watermelon than pounds of grapes are in the fruit salad? Show your work.

   $5/10$ or $1/2$

B. Keisha added $\frac{7}{10}$ pound strawberries to the fruit salad. How many more pounds of strawberries than pounds of blueberries are in the salad? Show your work.

   $7/10 - 2/10 =$ $5/10$ or $1/2$

   $5/10 = 1/2$

Add and Subtract Mixed Numbers

Common Core State Standards:
4.NF.3.b, 4.NF.3.c

Getting the Idea

A mixed number is made up of a whole number and a fraction.

You can show a mixed number as an equation.

$$2\frac{1}{4} = 1 + 1 + \frac{1}{4}$$

$$2\frac{1}{4} = \frac{4}{4} + \frac{4}{4} + \frac{1}{4}$$

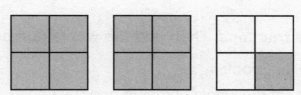

Example 1

Show the mixed number $1\frac{2}{5}$ as an equation of fractions.

Make a model to show the equation.

Strategy   **Break the mixed number into whole numbers and fractions.**

Step 1   Show the mixed number as whole numbers and fractions.

$$1\frac{2}{5} = 1 + \frac{2}{5}$$

Step 2   Show the 1 whole as a fraction with 5 as a denominator.

$$1 = \frac{5}{5}$$

$$1\frac{2}{5} = \frac{5}{5} + \frac{2}{5}$$

Step 3   Make a model.

Draw rectangles. Show each rectangle in fifths.

$$1\frac{2}{5} = \frac{5}{5} + \frac{2}{5}$$

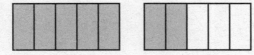

Solution $1\frac{2}{5} = \frac{5}{5} + \frac{2}{5}$

Remember that a unit fraction has a 1 in the numerator.

In Example 1, you could also use all unit fractions.

$$1\frac{2}{5} = \frac{1}{5} + \frac{1}{5} + \frac{1}{5} + \frac{1}{5} + \frac{1}{5} + \frac{1}{5} + \frac{1}{5}$$

When you add mixed numbers, first add the fraction parts, then add the whole number parts.

Example 2

Add.

$$1\frac{1}{4} + 2\frac{1}{4} = \boxed{}$$

Strategy   **Add the fractions. Then add the whole numbers.**

Step 1   Rewrite the problem.

Line up the whole numbers and fractions.

$$\begin{array}{r} 1\frac{1}{4} \\ + 2\frac{1}{4} \\ \hline \end{array}$$

Step 2   Add the fraction parts: $\frac{1}{4} + \frac{1}{4}$.

$$\begin{array}{r} 1\frac{1}{4} \\ + 2\frac{1}{4} \\ \hline \frac{2}{4} \end{array}$$

Step 3   Add the whole number parts: $1 + 2$.

$$\begin{array}{r} 1\frac{1}{4} \\ + 2\frac{1}{4} \\ \hline 3\frac{2}{4} \end{array}$$

Step 4   Simplify the fraction part.

$$\frac{2}{4} = \frac{2 \div 2}{4 \div 2} = \frac{1}{2}$$

So, $3\frac{2}{4} = 3\frac{1}{2}$.

Solution $1\frac{1}{4} + 2\frac{1}{4} = 3\frac{1}{2}$

Another way to add mixed numbers is to write each mixed number as an equivalent improper fraction. Then add the improper fractions. Show the answer as a mixed number.

Example 3

Kelsey drank $1\frac{2}{3}$ cups of water in the morning and $2\frac{2}{3}$ cups of water in the afternoon.

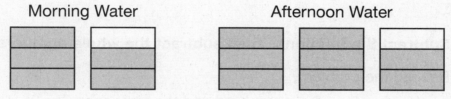

Morning Water   Afternoon Water

How much water did Kelsey drink in all?

Strategy **Write each mixed number as an improper fraction. Then add.**

Step 1 Write a number sentence for the problem.

Let w represent the amount of water Kelsey drank in all.

$1\frac{2}{3} + 2\frac{2}{3} = w$

Step 2 Write $1\frac{2}{3}$ as an equivalent improper fraction.

$1\frac{2}{3} = 1 + \frac{2}{3}$

$1 = \frac{3}{3}$, because $\frac{1}{3} + \frac{1}{3} + \frac{1}{3} = \frac{3}{3}$

$1\frac{2}{3} = \frac{3}{3} + \frac{2}{3} = \frac{5}{3}$

Step 3 Write $2\frac{2}{3}$ as an equivalent improper fraction.

$2\frac{2}{3} = 2 + \frac{2}{3}$

$2 = \frac{6}{3}$, because $\frac{3}{3} + \frac{3}{3} = \frac{6}{3}$

$2\frac{2}{3} = \frac{6}{3} + \frac{2}{3} = \frac{8}{3}$

Step 4 Add the improper fractions.

$\frac{5}{3} + \frac{8}{3} = \frac{13}{3}$

Step 5 Change the sum to a mixed number.

Divide the numerator by the denominator.

$13 \div 3 = 4\text{ R}1$

So, $\frac{13}{3} = 4\frac{1}{3}$.

Solution **Kelsey drank $4\frac{1}{3}$ cups of water in all.**

When you subtract mixed numbers, first subtract the fraction parts, then subtract the whole number parts.

Example 4

Subtract.

$$3\frac{5}{8} - 1\frac{3}{8} = \boxed{}$$

Strategy   **Subtract the fractions. Then subtract the whole numbers.**

Step 1   Rewrite the problem.

Line up the whole numbers and the fractions.

$$\begin{array}{r} 3\frac{5}{8} \\ -\ 1\frac{3}{8} \\ \hline \end{array}$$

Step 2   Subtract the fraction parts: $\frac{5}{8} - \frac{3}{8}$.

$$\begin{array}{r} 3\frac{5}{8} \\ -\ 1\frac{3}{8} \\ \hline \frac{2}{8} \end{array}$$

Step 3   Subtract the whole number parts: $3 - 1$.

$$\begin{array}{r} 3\frac{5}{8} \\ -\ 1\frac{3}{8} \\ \hline 2\frac{2}{8} \end{array}$$

Step 4   Simplify the fraction part.

$$\frac{2}{8} = \frac{2 \div 2}{8 \div 2} = \frac{1}{4}$$

So, $2\frac{2}{8} = 2\frac{1}{4}$.

Solution $3\frac{5}{8} - 1\frac{3}{8} = 2\frac{1}{4}$

You can use the relationship between addition and subtraction to find a missing number.

Example 5

What mixed number goes in the box to make the sentence true?

$$\boxed{} - 2\frac{1}{3} = 3\frac{1}{3}$$

Strategy   **Use the relationship between addition and subtraction.**

Step 1   Addition and subtraction are inverse operations.

To find $\boxed{}$, add $2\frac{1}{3} + 3\frac{1}{3}$.

Step 2   Add $2\frac{1}{3}$ and $3\frac{1}{3}$.

Add the fraction parts.

Then add the whole number parts.

$$
\begin{array}{r}
2\frac{1}{3} \\
+\ 3\frac{1}{3} \\
\hline
5\frac{2}{3}
\end{array}
$$

Step 3   Subtract to check your answer.

$$
\begin{array}{r}
5\frac{2}{3} \\
-\ 2\frac{1}{3} \\
\hline
3\frac{1}{3}
\end{array}
$$

Solution $5\frac{2}{3} - 2\frac{1}{3} = 3\frac{1}{3}$

Coached Example

Mr. Lee bought a $4\frac{1}{4}$-feet-long wooden board. He wants to cut a piece that is $2\frac{3}{4}$ feet from the board. How long will the board be that Mr. Lee has left?

Write a number sentence for the problem.

The board is _____ feet long.

He will cut a _____ feet piece.

Let b represent the length of board left.

_____ − _____ = _____

Write $4\frac{1}{4}$ as an equivalent improper fraction.

Write the whole number part as a fraction with a denominator of 4.

$4 = \frac{4}{4} +$ _____ + _____ + _____

Add the fractions.

$4\frac{1}{4} = \frac{4}{4} + \frac{4}{4} + \frac{4}{4} + \frac{4}{4} + \frac{1}{4} =$ _____

So, $4\frac{1}{4} =$ _____.

Write $2\frac{3}{4}$ as an equivalent improper fraction.

Write the whole number part as a fraction with a denominator of 4.

$2 =$ _____ + _____

Add the fractions.

$2\frac{3}{4} =$ _____ + _____ + _____ = _____

So, $2\frac{3}{4} =$ _____.

Subtract the improper fractions.

_____ − _____ = _____

Change the improper fraction to mixed number.

Divide the numerator by the denominator.

_____ ÷ _____ = _____

The mixed number is _____.

Simplify the fraction part of the mixed number. _____

Mr. Lee has _____ feet of the board left.

Lesson Practice

Choose the correct answer.

1. Which mixed number does this show?

$$\frac{3}{3} + \frac{3}{3} + \frac{1}{3} + \frac{1}{3}$$

   A. $2\frac{1}{3}$

   B. $1\frac{2}{3}$

   C. $2\frac{2}{3}$

   D. $3\frac{1}{3}$

2. Which equation is true?

   A. $2\frac{1}{12} = \frac{12}{12} + \frac{12}{12} + \frac{12}{12}$

   B. $2\frac{1}{12} = \frac{12}{12} + \frac{12}{12} + \frac{10}{12}$

   C. $2\frac{1}{12} = \frac{12}{12} + \frac{12}{12} + \frac{1}{12}$

   D. $2\frac{1}{12} = \frac{1}{12} + \frac{1}{12} + \frac{1}{12}$

3. Add.

$$3\frac{2}{5} + 3\frac{2}{5} = \boxed{}$$

   A. $4\frac{5}{6}$

   B. $5\frac{2}{5}$

   C. $6\frac{2}{5}$

   D. $6\frac{4}{5}$

4. Subtract.

$$4\frac{3}{4} - 2\frac{1}{4} = \boxed{2\frac{3}{4}}$$

   A. $2\frac{1}{2}$

   B. $2\frac{1}{4}$

   C. $6\frac{1}{2}$

   D. $7\frac{1}{4}$

5. Connor watched $13\frac{1}{3}$ hours of TV this week and $7\frac{1}{3}$ hours last week. How many hours of TV did Connor watch in all in 2 weeks?

   A. $6\frac{1}{3}$ hours

   B. $19\frac{1}{3}$ hours

   C. $20\frac{2}{3}$ hours

   D. $11\frac{2}{3}$ hours

6. Nora has a red ribbon that is $3\frac{1}{8}$ feet long. She also has a purple ribbon that is $1\frac{5}{8}$ feet long. How many feet longer is the red ribbon than the purple ribbon?

   A. $1\frac{1}{4}$ feet   C. $2\frac{1}{8}$ feet

   B. $1\frac{1}{2}$ feet   D. $3\frac{2}{3}$ feet

7. Which number goes in the box to make the sentence true?

$$5\frac{7}{10} - \boxed{} = 2\frac{3}{10}$$

A. $7\frac{2}{5}$

B. $3\frac{4}{5}$

C. $3\frac{2}{5}$

D. $3\frac{1}{4}$

8. Which number goes in the box to make the sentence true?

$$\boxed{} - 1\frac{1}{8} = 2\frac{5}{8}$$

A. $4\frac{3}{4}$

B. $3\frac{3}{4}$

C. $3\frac{1}{2}$

D. $1\frac{1}{2}$

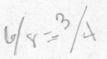

9. Gordon bought $6\frac{1}{4}$ pounds of ground beef to make food for a party. He used $2\frac{1}{4}$ pounds of beef to make lasagna and another $2\frac{1}{4}$ pounds to make hamburgers.

A. How many pounds of beef did Gordon use in all to make lasagna and hamburgers? Show your work.

2 1/4 · 2 1/4 = 4 2/4 = 4 1/2

4 2/4

B. How much ground beef does Gordon have left? Show your work.

2 pounds

6 1/4 - 4 1/4 = 2

Common Core State Standards:
4.NF.4.a, 4.NF.4.b, 4.NF.4.c

Multiply Fractions with Whole Numbers

Getting the Idea

You can use models to help you multiply a fraction by a whole number.

Example 1

Multiply.

$$4 \times \frac{1}{2} = \boxed{}$$

Strategy   **Use models.**

Step 1   Show 4 groups of $\frac{1}{2}$.

$$\frac{1}{2} \quad \frac{1}{2} \quad \frac{1}{2} \quad \frac{1}{2} \qquad = \qquad \frac{4}{2} = 2$$

Step 2   Find the product.

There are 4 halves, or 2 wholes, in all.

Solution $4 \times \frac{1}{2} = \frac{4}{2} = 2$

Multiplication is the same as repeated addition. You can use repeated addition to multiply a fraction by a whole number.

For example, to multiply $5 \times \frac{1}{4}$, add $\frac{1}{4}$ five times.

$$\frac{1}{4} + \frac{1}{4} + \frac{1}{4} + \frac{1}{4} + \frac{1}{4} \qquad = \qquad \frac{5}{4}$$

So, $5 \times \frac{1}{4} = \frac{5}{4}$.

Since $\frac{5}{4}$ is the product of 5 and $\frac{1}{4}$, $\frac{5}{4}$ is a multiple of $\frac{1}{4}$.

Example 2

Multiply.

$$3 \times \frac{2}{3} = \square$$

Strategy   **Use repeated addition.**

Step 1   Write the multiplication as repeated addition.

$$3 \times \frac{2}{3} = \frac{2}{3} + \frac{2}{3} + \frac{2}{3}$$

Step 2   Add the numerators. The denominator stays the same.

$$\frac{2}{3} + \frac{2}{3} + \frac{2}{3} = \frac{2+2+2}{3} = \frac{6}{3}$$

Step 3   Simplify.

$$\frac{6}{3} = 2$$

Step 4   Use models to check.

Show 3 groups of $\frac{2}{3}$.

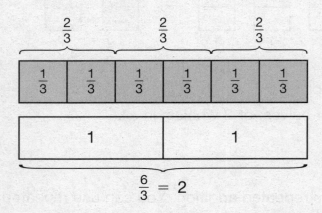

There are 2 groups of $\frac{3}{3}$.

$$\frac{3}{3} + \frac{3}{3} = 1 + 1$$

$$1 + 1 = 2$$

Solution $3 \times \frac{2}{3} = 2$

In Example 2, the product is $\frac{6}{3}$ or 2. So, 2 is a multiple of $\frac{2}{3}$.

You can also show $3 \times \frac{2}{3}$ as $6 \times \frac{1}{3}$. Both give the same product.

So, 2 is also a multiple of $\frac{1}{3}$.

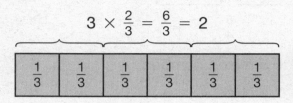

$$3 \times \frac{2}{3} = \frac{6}{3} = 2 \qquad\qquad 6 \times \frac{1}{3} = \frac{6}{3} = 2$$

When you multiply a fraction and a whole number, you can rename the whole number as a fraction. Then multiply the numerators and the denominators.

Example 3

Carrie and Eddie each ordered $\frac{3}{4}$ pound of chocolate. How much chocolate did they order in all? Between what two whole numbers does the answer lie?

Strategy   **Write the whole number as a fraction. Multiply.**

Step 1   Write the multiplication sentence for the problem.

Each person ordered $\frac{3}{4}$ pound.

Let t represent the total number of pounds of chocolate ordered.

$2 \times \frac{3}{4} = t$

Step 2   Write the whole number as a fraction.

$2 = \frac{2}{1}$

Step 3   Multiply the numerators and denominators.

$\frac{2}{1} \times \frac{3}{4} = \frac{2 \times 3}{1 \times 4} = \frac{6}{4}$

Step 4 Use models to check.

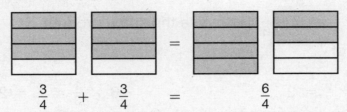

$$\frac{3}{4} \quad + \quad \frac{3}{4} \quad = \quad \frac{6}{4}$$

Step 5 Change the improper fraction to a mixed number.

$$\frac{6}{4} = \frac{4}{4} + \frac{2}{4}$$

$$\frac{4}{4} + \frac{2}{4} = 1 + \frac{2}{4}$$

$$1 + \frac{2}{4} = 1\frac{2}{4}$$

Step 6 Simplify the mixed number.

$$\frac{2}{4} = \frac{2 \div 2}{4 \div 2} = \frac{1}{2}$$

so $1\frac{2}{4} = 1\frac{1}{2}$.

Step 7 Use a number line to find the two whole numbers the sum lies between. Put a point at $1\frac{1}{2}$.

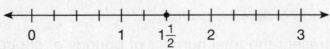

The sum is between 1 and 2.

Solution **They ordered $1\frac{1}{2}$ pounds of chocolate in all.**
The answer lies between 1 and 2.

Coached Example

Multiply.

$$4 \times \frac{3}{5} = \boxed{}$$

Write the multiplication as repeated addition.

$4 \times \frac{3}{5} =$ _____ + _____ + _____ + _____

Look at the denominators.

Are the denominators the same? _____

Add the numerators.

_____ + _____ + _____ + _____ = _____

Write the sum over the denominator. _____

Change the sum to a mixed number.

Divide the numerator by the denominator.

$\frac{12}{5} =$ _____

Make a model of the problem to check your answer.

$4 \times \frac{3}{5} =$ _____

Lesson Practice

Choose the correct answer.

1. Multiply.

$$5 \times \frac{1}{6} = \boxed{}$$

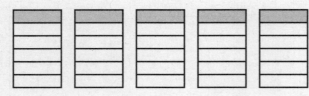

    A. $\frac{1}{3}$

    B. $\frac{1}{2}$

    C. $\frac{2}{3}$

    D. $\frac{5}{6}$

2. Multiply.

$$3 \times \frac{2}{10} = \boxed{}$$

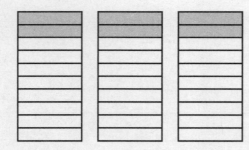

    A. $\frac{5}{10}$

    B. $\frac{6}{10}$

    C. $\frac{8}{7}$

    D. 1

3. Which will have the same product as $4 \times \frac{2}{5}$?

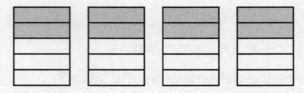

    A. $8 \times \frac{1}{5}$

    B. $4 \times \frac{1}{2}$

    C. $5 \times \frac{1}{4}$

    D. $4 \times \frac{1}{5}$

4. Multiply.

$$8 \times \frac{1}{4} = \boxed{}$$

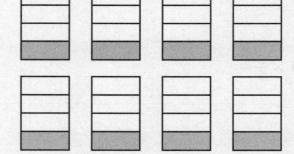

    A. $\frac{4}{8}$

    B. $\frac{1}{2}$

    C. 1

    D. 2

5. Multiply.

$$3 \times \frac{1}{12} = \boxed{}$$

A. $\frac{2}{12}$ **C.** $\frac{4}{12}$

B. $\frac{3}{12}$ **D.** 4

6. Multiply.

$$4 \times \frac{1}{2} = \boxed{}$$

A. 1

B. $\frac{3}{2}$

C. 2

D. $\frac{5}{2}$

7. Saul put $\frac{1}{2}$ bag of pretzels in his lunch box each day for 3 days. How many bags of pretzels did Saul put in his lunch box in all?

A. $\frac{2}{3}$ **C.** 2

B. $1\frac{1}{2}$ **D.** $2\frac{1}{2}$

8. Catalina decorated each of 5 shoeboxes with $\frac{1}{5}$ foot of ribbon. How many feet of ribbon did she use in all?

A. $\frac{2}{5}$ foot   **C.** $\frac{4}{5}$ foot

B. $\frac{3}{5}$ foot   **D.** 1 foot

9. Theresa has 9 baskets. Each basket is $\frac{1}{3}$ full of tomatoes.

    A. Make a model to show the problem.

    B. How many full baskets of tomatoes does Theresa have? Use multiplication. Show your work.

Decimals

Common Core State Standard:
4.NF.6

Getting the Idea

A **decimal** can name part of a whole or part of a group. A decimal can have a whole number part and a decimal part that are separated by a **decimal point (.)**.

Each grid below represents 1 whole. The decimal shows the shaded part of the whole.

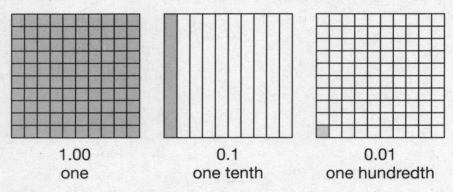

| 1.00 | 0.1 | 0.01 |
| one | one tenth | one hundredth |

Place values in decimals follow the same base-ten system as whole numbers.

Each place has 10 times the value of the place to its right.

    1 tenth = 10 hundredths

    1 one = 10 tenths

When you put a 0 to the right of the last digit of a decimal, it does not change the value of the decimal.

For example, 0.1 = 0.10 and 0.5 = 0.50.

You can use a place-value chart to show the value of each digit in a decimal. The decimal 2.35 has 2 ones, 3 tenths, and 5 hundredths, or 2 ones and 35 hundredths.

Ones	.	Tenths	Hundredths
2	.	3	5

Example 1

What decimal names the shaded part of the grid?

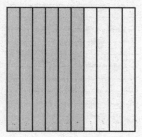

Strategy	**The entire grid equals 1 whole. Find the decimal for each part.**

Step 1   Find the decimal that represents each part of the grid.

There are 10 parts in the grid.

Each part is 0.1 or one tenth.

Step 2   Count the number of shaded parts.

There are 6 shaded parts.

So, six tenths, or 0.6, of the grid is shaded.

Solution   **The decimal 0.6 names the shaded part of the grid.**

To read or write the word name of a decimal less than 1, read the number to the right of the decimal point as you would read a whole number. Then read the least place value.

0.4 is read as *four tenths*.

0.23 is read as *twenty-three hundredths*.

To read or write the word name of a decimal greater than 1, read the whole number, use the word *and* for the decimal point, and then read the decimal part.

3.6 is read as *three and six tenths*.

1.27 is read as *one and twenty-seven hundredths*.

Example 2

What decimal names the shaded part of the grid?

Strategy   **The entire grid equals 1 whole. Find the decimal for each part.**

Step 1   Find the decimal that represents each part of the grid.

There are 100 parts in the grid.

Each part is 0.01 or one hundredth.

Step 2   Count the number of shaded parts.

There are 57 shaded parts.

Step 3   Write the decimal in a place-value chart.

Ones	.	Tenths	Hundredths
0	.	5	7

Solution   **The decimal 0.57, or fifty-seven hundredths, names the shaded part of the grid.**

Example 3

What decimal do the models show? What is the word name for the decimal?

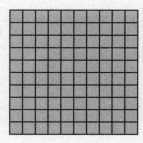 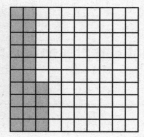

Strategy	**Use a place-value chart.**

Step 1   Find the decimal represented by the models.

There is 1 whole grid shaded.

The other grid has 24 out of 100 parts shaded.

So, 0.24 of the other grid is shaded.

Step 2   Write the decimal in a place-value chart.

Ones	.	Tenths	Hundredths
1	.	2	4

Step 3   Write the part to the left of the decimal point in words.

The whole number part is *one*.

The decimal point is *and*.

Step 4   Write the part to the right of the decimal point in words.

The decimal part is *twenty-four*.

The least place value is *hundredths*.

Solution   **The models show the decimal 1.24.**
The word name is *one and twenty-four hundredths*.

You can represent decimals on a number line. Count the equal parts between marked numbers to decide what each tick mark represents.

Example 4

What decimal does point *H* represent on the number line?

Strategy   **Decide what each tick mark represents.**

Step 1   Count the equal parts between marked numbers.

There are 10 equal parts between 1 and 2.

So, each tick mark represents 0.1 or one tenth.

Step 2   Count the spaces from 1 to point *H*.

Point *H* is at the fourth mark after 1.

Since each mark represents 0.1, point *H* is at 1.4.

Solution   **Point *H* represents 1.4.**

Example 5

What decimal does point *J* represent on the number line?

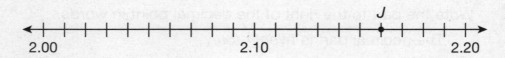

Strategy   **Decide what each tick mark represents.**

Step 1   Count the equal parts between marked numbers.

There are 10 equal parts between 2.10 and 2.20.

So, each tick mark represents 0.01 or one hundredth.

Step 2   Count the spaces from 2.10 to point *J*.

Point *J* is at the sixth mark after 2.10.

Since each mark represents 0.01, point *J* is at 2.16.

Solution   **Point *J* represents 2.16.**

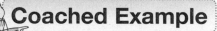 **Coached Example**

What decimal do the models show? What is the word name for the decimal?

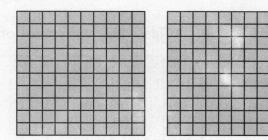 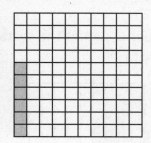

There are _____ whole grids shaded.

The other grid has _____ out of _____ parts shaded.

So, _____ of the other grid is shaded.

Write the decimal in a place-value chart.

Ones	.	Tenths	Hundredths

Write the part to the left of the decimal point in words.

The whole number part is _____.

The decimal point is _____.

Write the part to the right of the decimal point in words.

The decimal part is _____.

The least place value is _____.

The models show the decimal _____.

The word name is _____.

Lesson Practice

Choose the correct answer.

1. What decimal names the shaded part of the grid?

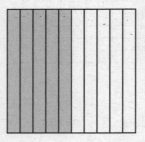

   A. 0.01

   B. 0.05

   C. 0.1

   D. 0.5

2. What decimal names the shaded part of the grid?

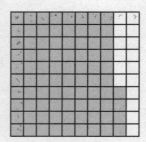

   A. 0.16

   B. 0.40

   C. 0.84

   D. 0.94

3. What is the word name of the decimal?

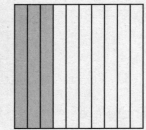

   A. three tenths

   B. one and three tenths

   C. one and three hundredths

   D. three and one tenth

4. How many hundredths are in the decimal 1.73?

   A. 1   **C.** 4

   B. 3   **D.** 7

5. Which decimal is three and nine hundredths?

   A. 3.90   **C.** 0.93

   B. 3.09   **D.** 0.39

6. Which decimal has an 8 in the tenths place?

   A. 1.84   **C.** 8.19

   B. 3.28   **D.** 80.25

Use the number line for questions 7 and 8.

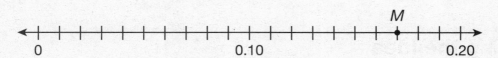

7. What decimal does point M represent on the number line?

    A. 0.07

    B. 0.17

    C. 0.7

    D. 1.7

8. How can you read the decimal that point M represents?

    A. seven hundredths

    B. seventeen tenths

    C. seventeen hundredths

    D. one and seven tenths

9. Mr. Tyler drew these grids on the board.

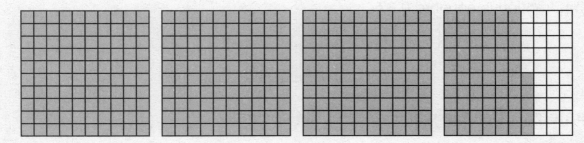

    A. What decimal do the grids show?

    _____

    B. How many ones, tenths, and hundredths are in the decimal?

    _____

    C. How can you write the word name of the decimal?

    _____

_____

Common Core State Standards:
4.NF.5, 4.NF.6

Relate Decimals to Fractions

Getting the Idea

Decimals and fractions are related. Both can name part of a whole.
A decimal in tenths can be written as a fraction with a denominator of 10.
A decimal in hundredths can be written as a fraction with a denominator of 100.

Each grid below represents 1 whole. The decimal and fraction both name the shaded part of each grid.

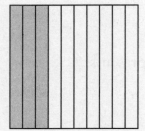 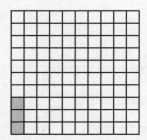

3 shaded parts out of 10 equal parts   3 shaded parts out of 100 equal parts

$0.3 = $ three tenths $= \frac{3}{10}$ $0.03 = $ three hundredths $= \frac{3}{100}$

Example 1

What fraction is equal to 0.8?

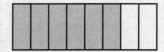

Strategy   **Use the place value of the decimal.**

Step 1   Find the place value of the decimal.

The decimal shows 8 tenths.

Step 2   Write the fraction.

The denominator is 10. The numerator is 8.

$\frac{8}{10}$

Solution $0.8 = \frac{8}{10}$

Example 2

What fraction is equal to 0.07?

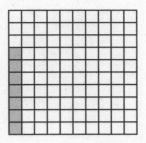

Strategy   **Use the place value of the decimal.**

Step 1   Find the place value of the decimal.

The decimal shows 7 hundredths.

Step 2   Write the fraction.

The denominator is 100. The numerator is 7.

$$\frac{7}{100}$$

Solution $0.07 = \frac{7}{100}$

Example 3

In simplest form, what fraction is equal to 0.4?

Strategy   **Read the decimal to change a decimal to a fraction. Then simplify.**

Step 1   Read the decimal. Write the fraction.

$0.4 = 4 \text{ tenths} = \frac{4}{10}$

Step 2   Simplify the fraction.

The greatest common factor of 4 and 10 is 2.

$$\frac{4}{10} = \frac{4 \div 2}{10 \div 2} = \frac{2}{5}$$

Solution   **In simplest form, the fraction $\frac{2}{5}$ is equal to 0.4.**

You can express a fraction with a denominator of 10 as an equivalent fraction with a denominator of 100. Multiply the numerator and denominator by 10.

For example,

$$\frac{9}{10} = \frac{9 \times 10}{10 \times 10} = \frac{90}{100}$$

Example 4

The shaded part of the grid shows 0.79.

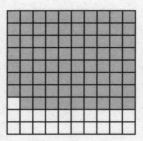

What fraction is equal to that decimal? Write the fraction as an equation.

Strategy   **Break the shaded part into tenths and hundredths.**

Step 1   Show the decimal in tenths and hundredths.

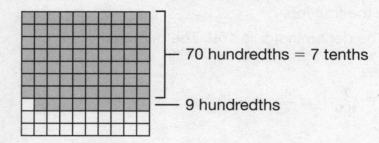

70 hundredths = 7 tenths

9 hundredths

0.79 = 79 hundredths = 7 tenths 9 hundredths

Step 2   Show the parts as fractions.

7 tenths = $\frac{7}{10}$

9 hundredths = $\frac{9}{100}$

So, 79 hundredths = $\frac{7}{10} + \frac{9}{100}$.

Step 3   Show $\frac{7}{10}$ as an equivalent fraction with a denominator of 100.

$$\frac{7}{10} = \frac{7 \times 10}{10 \times 10} = \frac{70}{100}$$

Step 4   Write the decimal as a sum of two fractions.

$$0.79 = 79 \text{ hundredths} = \frac{79}{100}$$

$$\frac{79}{100} = \frac{70}{100} + \frac{9}{100}$$

Solution   **The fraction $\frac{79}{100}$ is equal to 0.79.**

$$\frac{79}{100} = \frac{70}{100} + \frac{9}{100}$$

$$\frac{79}{100} = \frac{7}{10} + \frac{9}{100}$$

Coached Example

The shaded part of the grid shows 0.56.

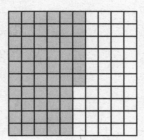

What fraction is equal to that decimal? Write the fraction as an equation.

Show the decimal in tenths and hundredths.

0.56 = _____ hundredths = _____ tenths _____ hundredths

Show the parts as fractions.

5 tenths = _____

6 hundredths = _____

So, 56 hundredths = _____ + _____.

Show $\frac{5}{10}$ as an equivalent fraction with a denominator of 100.

$\frac{5}{10}$ = _____

Write the decimal as a sum of two fractions.

0.56 = 56 hundredths = _____

$\frac{56}{100}$ = _____ + _____

The fraction _____ is equal 0.56.

$\frac{56}{100} = \frac{}{100} + \frac{}{100}$

$\frac{56}{100} = \frac{}{10} + \frac{}{100}$

Lesson Practice

Choose the correct answer.

1. What fraction is equal to 0.9?

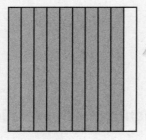

- **A.** $\frac{1}{9}$
- **B.** $\frac{9}{10}$
- **C.** $\frac{9}{100}$
- **D.** $\frac{90}{10}$

2. What fraction is equal to 0.04?

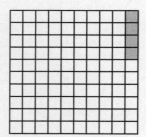

- **A.** $\frac{1}{4}$
- **B.** $\frac{4}{10}$
- **C.** $\frac{4}{100}$
- **D.** $\frac{40}{10}$

3. Rod's puppy ate 8 tenths of a can of dog food. Which model shows the amount of food the puppy ate?

A.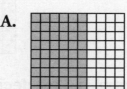

B.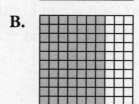

C.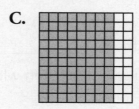

D.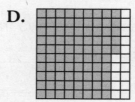

4. What fraction is equal to 0.19?

- **A.** $\frac{100}{19}$
- **C.** $\frac{19}{10}$
- **B.** $\frac{19}{100}$
- **D.** $\frac{1}{19}$

5. In simplest form, which fraction is equal to 0.5?

   A. $\frac{5}{100}$ **C.** $\frac{1}{5}$

   B. $\frac{10}{5}$ **D.** $\frac{1}{2}$

6. Which does **not** show the shaded part of the grid?

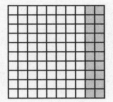

   A. $\frac{1}{10}$

   B. $\frac{1}{5}$

   C. $\frac{2}{10}$

   D. $\frac{20}{100}$

7. Which has a different value than all of the others?

   A. $\frac{9}{100}$ **C.** 0.9

   B. $\frac{9}{10}$ **D.** 0.90

8. The grid has 0.38 shaded.

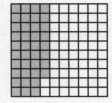

Which sum of fractions shows 0.38?

   A. $\frac{3}{10} + \frac{8}{10}$

   B. $\frac{3}{10} + \frac{8}{100}$

   C. $\frac{3}{100} + \frac{8}{100}$

   D. $\frac{3}{100} + \frac{8}{10}$

9. Lamar shaded a grid.

   A. Write the decimal for the shaded part.
Write the fraction in simplest form for the shaded part.

   _____

   _____

   B. Show the decimal as a sum of two fractions.

   _____

Compare and Order Decimals

Common Core State Standard:
4.NF.7

Getting the Idea

Remember to refer to the same whole when comparing decimals.

Use the same symbols to compare decimals.

The symbol $>$ means *is greater than*.

The symbol $<$ means *is less than*.

The symbol $=$ means *is equal to*.

Example 1

What symbol makes this sentence true? Write $>$, $<$, or $=$.

0.62 $\bigcirc$ 0.57

Strategy **Make a model for each decimal.**

Step 1 Use 10-by-10 grids. Shade the squares to show each decimal.

Each grid represents 1 whole.

0.62 = 62 hundredths   0.57 = 57 hundredths

   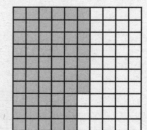

Step 2 Compare the shaded parts.

0.62 has more shaded parts.

0.62 is greater than 0.57.

Step 3 Choose the correct symbol.

$>$ means *is greater than*.

Solution 0.62 $\left(>\right)$ 0.57

You can use a place-value chart to help compare decimals.
Start comparing the digits in the greatest place.

Example 2

What symbol makes this sentence true? Write >, <, or =.

$$0.76 \bigcirc 0.74$$

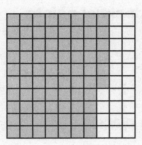 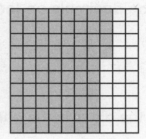

Strategy   Use a place-value chart. Start with the greatest place.

Ones	·	Tenths	Hundredths
0	·	7	6
0	·	7	4

Step 1   Compare the digits in the ones place.

0 ones = 0 ones

Compare the next greatest place.

Step 2   Compare the digits in the tenths place.

7 tenths = 7 tenths

Compare the next greatest place.

Step 3   Compare the digits in the hundredths place.

6 hundredths > 4 hundredths

0.76 is greater than 0.74.

Step 4   Choose the correct symbol.

> means *is greater than*.

Solution   0.76 $\bigcirc\!\!>$ 0.74

When using place value to compare, line up the decimals on the decimal points. Compare the place values from left to right.

Example 3

List the decimals below in order from greatest to least.

0.81   1.03   0.88

Strategy   **Line up the decimals on the decimal points.**

0.81
1.03
0.88

Step 1   Compare the digits in the ones place.

0.81
1.03
0.88

$1 > 0$, so 1.03 is the greatest decimal.

Step 2   Compare the digits in the tenths place.

0.**8**1
0.**8**8

$8 = 8$, so compare the next place.

Step 3   Compare the digits in the hundredths place.

0.8**1**
0.8**8**

$1 < 8$, so 0.81 is the least decimal.

Solution   **From greatest to least, the order of the decimals is 1.03, 0.88, 0.81.**

Coached Example

Order the decimals below from least to greatest.

0.40   **0.52**   **0.48**

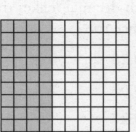  

Use place value to order the decimals.

Write the decimals in a place-value chart.

Ones	·	Tenths	Hundredths
			.

Compare the digits in the greatest place, the _____.

_____ ones = _____ ones = _____ ones

All of the digits in the greatest place are _____.

Compare the digits in the next greatest place, the _____.

_____ tenths > _____ tenths, so _____ is the greatest decimal.

Compare the remaining two decimals.

Compare the digits in the next greatest place, the _____.

_____ hundredths < _____ hundredths, so _____ is the least decimal.

From least to greatest, the order of the decimals is

_____, _____, _____.

Lesson Practice

Choose the correct answer.

1. What symbol makes this sentence true?

    0.38 ◯ 0.36

    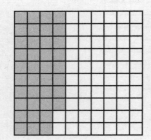

    A. >

    B. <

    C. =

    D. +

2. Which sentence is true?

    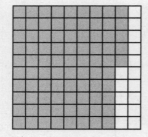 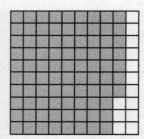

    0.85   0.87

    A. 0.85 = 0.87

    B. 0.87 < 0.85

    C. 0.85 > 0.87

    D. 0.87 > 0.85

3. Which decimal makes this sentence true?

    0.28 < ☐

    A. 0.03

    B. 0.27

    C. 0.28

    D. 0.30

4. Which decimal is less than 3.83?

    A. 3.08

    B. 3.83

    C. 4.08

    D. 4.80

5. Which list is in order from greatest to least?

     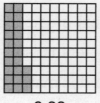 

    0.23   0.42   0.22

    A. 0.23   0.42   0.22

    B. 0.22   0.23   0.42

    C. 0.42   0.23   0.22

    D. 0.22   0.42   0.23

6. Which is the greatest decimal?

   A. 2.02   **C.** 1.75

   B. 0.99   **D.** 1.68

7. Which list is in order from greatest to least?

   A. 0.97   1.62   1.68

   B. 1.62   1.68   0.97

   C. 1.68   1.62   0.97

   D. 0.97   1.68   1.62

8. Which list is in order from least to greatest?

   A. 0.44   0.34   0.30

   B. 0.30   0.44   0.34

   C. 0.44   0.30   0.34

   D. 0.30   0.34   0.44

9. Joseph bought 0.78 pound of roast beef. He also bought 0.52 pound of provolone cheese.

   A. Shade the grids below to show each decimal.

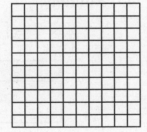 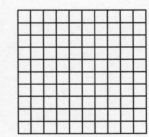

   B. Which item did Joseph buy more of, roast beef or provolone cheese?

_____

Domain 3: Cumulative Assessment for Lessons 18–27

1. Jessie has $\frac{8}{12}$ pound of cheese.

Which shows an equivalent fraction to $\frac{8}{12}$?

A.

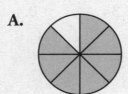

B.

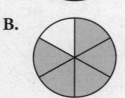

C.

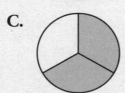

D.

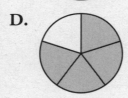

2. Cruz poured $\frac{2}{8}$ cup of milk into an empty pot. He then poured another $\frac{4}{8}$ cup of milk into the same pot. How much milk in all is in the pot?

A. $\frac{3}{8}$ cup

B. $\frac{5}{8}$ cup

C. $\frac{6}{8}$ cup

D. $\frac{7}{8}$ cup

3. In Nadine's class, $\frac{2}{8}$ of the students have summer birthdays. One-eighth of the students have autumn birthdays. What fraction more of Nadine's class has summer birthdays than autumn birthdays?

A. $\frac{0}{4}$

B. $\frac{1}{8}$

C. $\frac{1}{4}$

D. $\frac{3}{8}$

4. Multiply.

$$4 \times \frac{2}{3} = \boxed{}$$

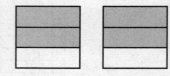
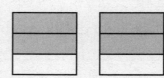

A. $\frac{2}{3}$

B. $1\frac{2}{3}$

C. 2

D. $2\frac{2}{3}$

5. Which sentence is true?

  A. $\frac{2}{3} > \frac{4}{6}$

  B. $\frac{3}{6} < \frac{5}{12}$

  C. $\frac{2}{5} < \frac{3}{10}$

  D. $\frac{5}{8} < \frac{3}{4}$

6. Which decimal represents the shaded part of the grids?

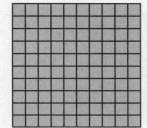 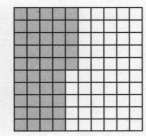

  A. 0.45

  B. 1.45

  C. 14.5

  D. 145

7. This grid has 0.77 shaded.

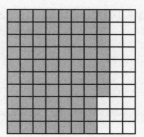

Which sum of fractions shows 0.77?

  A. $\frac{7}{10} + \frac{7}{10}$

  B. $\frac{7}{100} + \frac{7}{100}$

  C. $\frac{7}{10} + \frac{7}{100}$

  D. $\frac{7}{1} + \frac{7}{10}$

8. Which decimal makes the sentence true?

  $0.17 < \boxed{}$

  A. 0.09

  B. 0.15

  C. 0.17

  D. 0.20

9. Add.

  $2\frac{1}{6} + 3\frac{1}{6} = \boxed{}$

10. There are $3\frac{1}{4}$ pints of ice cream in the freezer. Tori ate $\frac{3}{4}$ pint.

  A. How much ice cream is left? Show your work.

  _____

  _____

  _____

  B. Show your answer for Part A in simplest form.

  _____

  _____

  _____

Domain 4

Measurement and Data

Domain 4: Diagnostic Assessment for Lessons 28–36

Domain 4: Cumulative Assessment for Lessons 28–36

Domain 4: Diagnostic Assessment for Lessons 28–36

1. The point on the number line below shows the height of a table.

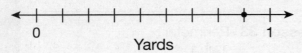

Yards

What is the height of the table?

A. $\frac{7}{8}$ yard

B. $\frac{7}{9}$ yard

C. $\frac{8}{9}$ yard

D. $\frac{9}{10}$ yard

2. Ebony started reading at 10:40 A.M. She stopped reading at 12:30 P.M. How long did Ebony spend reading?

A. 1 hour 30 minutes

B. 1 hour 50 minutes

C. 2 hours 30 minutes

D. 2 hours 50 minutes

3. A square has an area of 49 square inches. What is the length of one side of the square?

A. 6 inches

B. 7 inches

C. 9 inches

D. 11 inches

4. A chef needs 5 pounds of carrots for a salad. How many ounces are in 5 pounds?

A. 16 ounces   C. 64 ounces

B. 48 ounces   D. 80 ounces

5. A bathroom sink has a capacity of 4 liters. A bathtub has a capacity that is 50 times more than the sink. What is the capacity of the tub?

A. 54 liters   C. 240 liters

B. 200 liters   D. 450 liters

6. The area of this rectangle is 56 square centimeters.

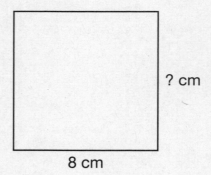

? cm

8 cm

What is the width of the rectangle?

A. 6 centimeters

B. 7 centimeters

C. 8 centimeters

D. 9 centimeters

7. The measure of angle *F* is 130°. A part of angle *F* measures 70°.

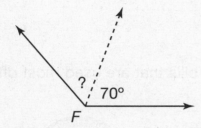

What is the measure of the other part of the angle?

A. 45°

B. 50°

C. 60°

D. 70°

8. What is the measure of this angle?

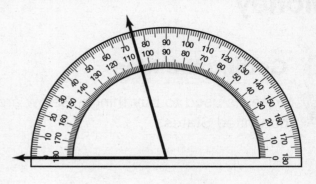

_____

9. Jamie recorded the number of miles he walked each day for a week. He made the line plot below.

Distance Walked

```
                X
                X
                X        X
  X             X        X
  |-------------|--------|-->
  0             1        1
                2
         Miles
```

A. How many days did Jamie walk $\frac{1}{2}$ mile?

_____

B. What is the difference in miles between the longest distance and the shortest distance Jamie walked?

_____

Money

Common Core State Standard:
4.MD.2

Getting the Idea

Money is used to buy things. Below are coins and bills that are used most often in the United States.

Penny	**Nickel**	**Dime**	**Quarter**	**Half Dollar**
1¢	5¢	10¢	25¢	50¢
$0.01	$0.05	$0.10	$0.25	$0.50

Dollar	**Five Dollars**	**Ten Dollars**	**Twenty Dollars**
$1.00	$5.00	$10.00	$20.00

To find the value of a group of bills and coins, count from the bill with the greatest value to the coin with the least value.

Example 1

Eda has these bills and coins.

How much money does Eda have?

Strategy   **Start with the bill with the greatest value.**

Step 1 Count the bills from the greatest value to least value.

$20.00 → $25.00 → $26.00 → $27.00

Step 2 Count the coins from the greatest value to the least value.
Continue from $27.00.

$27.25 → $27.50 → $27.60 → $27.65

Solution **Eda has $27.65.**

When you buy something at a store, you may not have the exact amount of money. You often give the cashier more money than the cost of your purchase and then get back change.

One way to make change is to count up from the price of the item to the amount of money you gave to the cashier.

Example 2

DeShawn bought a book for $8.59. He gave the cashier a $10 bill.
How much change should DeShawn receive from the cashier?

Strategy **Count up from the price of the book to the amount given.**

Step 1 Start with the price of the book and count up to $10.

$8.59 → $8.60 → $8.65 → $8.75 → $9.00 → $10.00

Step 2 Count the coins and bill to find the value of the change.

$1.00   → $1.25 → $1.35 → $1.40 → $1.41

Solution **DeShawn should receive $1.41 in change.**

You can also make change by subtracting the amount of a purchase from the amount given to the cashier.

Subtracting money amounts is similar to subtracting whole numbers, except that money amounts have decimal points. Line up the decimal points and subtract as you would with whole numbers. Be sure to include the decimal point and the dollar sign in the difference.

$$
\begin{array}{r}
\overset{9\ 9}{\overset{0\ \cancel{10}\ \cancel{10}\ 10}{\$\cancel{1}\,0.\cancel{0}\ \cancel{0}}} \\
-\ \$\ \ 8.5\ 9 \\
\hline
\$\ \ \ 1.4\ 1
\end{array}
$$

Example 3

Lana bought a sweater for $27.36. She gave the cashier $30.00. How much change should Lana receive?

Strategy   **Subtract as you would with whole numbers.**

Step 1   Write the number sentence for the problem.

Lana used $30.00 to buy a sweater that costs $27.36.

Let c represent the change Lana should receive.

$30.00 − $27.36 = c

Step 2   Set up the problem.

$$
\begin{array}{r}
\$30.00 \\
-\ \$27.36 \\
\hline
\end{array}
$$

Line up the decimal points.

Step 3   Subtract from right to left. Regroup if necessary.

$$
\begin{array}{r}
\overset{9\ 9}{\overset{2\ \cancel{10}\ \cancel{10}\ 10}{\$\cancel{3}\,0.\cancel{0}\,\cancel{0}}} \\
-\ \$27.36 \\
\hline
\$\ \ 2.64
\end{array}
$$

Solution   **Lana should receive $2.64 in change.**

Example 4

Kim has 4 library books that are all one day overdue. The library charges a late fee of $0.05 for each day a book is overdue. How much will Kim have to pay for the late fees?

Strategy   **Decide which operation to use.**

Step 1   Write the number sentence for the problem.

Kim has 4 books. Each book will be charged 5 cents.

Let f represent the total late fees.

$5 + 5 + 5 + 5 = f$ or $4 \times 5 = f$

Step 2   Multiply.

$4 \times 5 = 20$ cents

Step 3   Write the amount.

20 cents = $0.20

Solution   **Kim will have to pay $0.20 in late fees.**

Coached Example

Marvin has $20.00. He bought 2 books that cost $6.00 each and a bookmark that cost $1.50. How much money does Marvin have left?

Decide how to solve the problem.

Find the total amount Marvin spent.

Then subtract that amount from _____.

Marvin bought 2 books that cost _____ each.

So, 2 books cost _____.

Add the cost of 2 books to the cost of the bookmark.

_____ + _____ = _____

Marvin spent _____ in all.

Subtract the total amount from the amount Marvin has.

_____ − _____ = _____

Show your work.

Marvin has $_____ left.

Lesson Practice

Choose the correct answer.

1. How much money is shown below?

   A. $16.30

   B. $20.30

   C. $21.30

   D. $21.50

2. Rico's dinner cost $6.35. He paid with a $10 bill. How much change should Rico receive?

   A. $3.65

   B. $3.75

   C. $4.65

   D. $4.75

3. Marcy bought a magazine for $3.75 and a carton of juice for $2.25. She paid with a $10 bill. How much change should Marcy receive?

   A. $16.00

   B. $6.00

   C. $5.00

   D. $4.00

4. George bought 7 erasers. Each eraser costs $0.25. How much money did George spend?

   A. $1.25

   B. $1.75

   C. $2.75

   D. $7.25

5. Elizabeth had $50.00. She bought a shirt for $12.50 and a pair of pants for $24.00. How much money does Elizabeth have left?

   A. $13.50

   B. $14.50

   C. $26.00

   D. $36.50

6. Angie makes $5.00 each hour. She worked for 3 hours yesterday. How much money did Angie make yesterday?

   A. $8.00

   B. $8.30

   C. $10.00

   D. $15.00

7. Peter spent a total of $3.00 on 6 pencils. Each pencil cost the same amount. What did 1 pencil cost?

   A. $0.50

   B. $0.60

   C. $0.75

   D. $1.25

8. Erin needs to buy some school supplies. She has $5.00 to spend on supplies. Prices for supplies are shown on the price list below.

School Supplies Price List

Item	Price
Pen	$2.50
Pencil	$1.20
Eraser	$0.50
Ruler	$2.25

   A. Erin wants to buy a pen, a pencil, and 2 erasers. How much money will Erin spend in all? Show your work.

   B. How much money will Erin have left? Show your work.

Time

Common Core State Standards:
4.MD.1, 4.MD.2

Getting the Idea

Elapsed time is the amount of time that has passed from a beginning time to an end time. To find the elapsed time, count the **hours (hr)** then the **minutes (min)**.

Example 1

Shawna started and finished her homework at the times shown on the clocks.

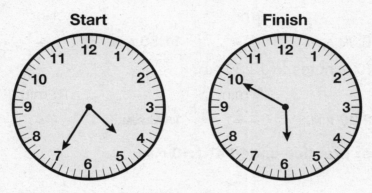

How long did it take Shawna to do her homework?

Strategy   **Find the elapsed time.**

Step 1   Read the start and finish times on the clocks.

    The start time is 4:35. The finish time is 5:50.

Step 2   Count the hours.

                    1 hr

    4:35   →   **5:35**

    There is one hour from 4:35 to 5:35.

Step 3   Count the minutes.

                    15 min

    5:35   →   **5:50**

    There are 15 minutes from 5:35 to 5:50.

Step 4   Add the times.

    1 hr + 15 min = 1 hr 15 min

Solution   **Shawna took 1 hour 15 minutes to do her homework.**

The time from midnight to noon is called A.M.
The time from noon to midnight is called P.M.

Example 2

Soccer practice began at 10:30 A.M. It lasted 2 hours 40 minutes.
At what time did soccer practice end?

Strategy   **Add the hours and minutes.**

Step 1   Add 2 hours to 10:30 A.M.

		1 hr				1 hr		
10:30 A.M.		→		**11:30 A.M.**		→		**12:30 P.M.**

Step 2   Add 40 minutes to 12:30 P.M.

		30 min				10 min		
12:30 P.M.		→		**1:00 P.M.**		→		**1:10 P.M.**

Solution   **Soccer practice ended at 1:10 P.M.**

The table shows the relationships of some units of time.

Units of Time
1 minute (min) = 60 seconds (sec)
1 hour (hr) = 60 minutes
1 **day (d)** = 24 hours
1 **week (wk)** = 7 days
1 **year (yr)** = 12 **months (mo)**

When you change from a larger unit to a smaller unit, use multiplication or addition.
When you change from a smaller unit to a larger unit, use division or subtraction.

Example 3

Emily slept for 8 hours last night. How many minutes did Emily sleep last night?

Strategy   **Use multiplication to change from hours to minutes.**

Step 1   Find the relationship between hours and minutes.

1 hour = 60 minutes

Step 2   Multiply 8 hours by 60 minutes.

$8 \times 60 = 480$ minutes

So 8 hours = 480 minutes.

Solution   **Emily slept for 480 minutes.**

Example 4

The table shows the relationship between the number of days and the number of weeks. Complete the table.

Week	Days
1	7
2	14
3	
4	
5	

Strategy   **Use the relationship between weeks and days.**

Step 1   Find the relationship between weeks and days.

1 week = 7 days

Step 2   Multiply 2 weeks by 7 days.

$2 \times 7 = 14$ days

2 weeks = 14 days

Step 3   Find the number of days in 3 to 5 weeks.

Multiply the number of weeks by 7.

$3 \times 7 = 21$ days

$4 \times 7 = 28$ days

$5 \times 7 = 35$ days

Complete the table.

Week	Days
1	7
2	14
3	21
4	28
5	35

Solution   The table is shown in Step 4.

Some word problems use fractions to show amounts of time. The table shows the relationships between hours and minutes.

Units of Time
$\frac{1}{4}$ hour = 15 minutes
$\frac{1}{2}$ hour = 30 minutes
$\frac{3}{4}$ hour = 45 minutes
1 hour = 60 seconds

Example 5

Malia spent a total of 2 hours in her workout. She spent $\frac{1}{4}$ of the workout time doing yoga. How many minutes did she spend doing yoga?

Strategy   Use multiplication.

Step 1   Write a number sentence for the problem.

She spent 2 hours in all.

She spent $\frac{1}{4}$ of the time doing yoga.

Let y represent the amount of time for yoga.

Find $\frac{1}{4} \times 2 = y$

Step 2   Multiply.

$$\frac{1}{4} \times 2 \text{ hr} = \frac{1}{4} \times \frac{2}{1} = \frac{1 \times 2}{4 \times 1} = \frac{2}{4} \text{ hr}$$

Step 3   Simplify $\frac{2}{4}$.

$$\frac{2}{4} = \frac{2 \div 2}{4 \div 2} = \frac{1}{2} \text{ hr}$$

Step 4   Find the number of minutes.

   She spent $\frac{1}{2}$ hour doing yoga.

   1 hour = 60 min, so $\frac{1}{2}$ hour = 30 min.

Solution   **Malia spent 30 minutes doing yoga.**

You can show intervals of time on a number line.

Example 6

Use the information in Example 5.
Show the time, in hours, that Malia spent doing yoga on a number line.

Strategy   **Label a number line in halves.**

Step 1   Draw a number line. Label it in halves.

   Label the units.

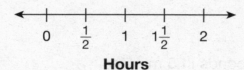

Hours

Step 2   Label the time that Malia spent doing yoga.

   She spent $\frac{1}{2}$ hour doing yoga.

   So label a point at $\frac{1}{2}$ on the number line.

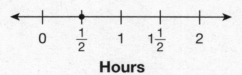

Hours

Solution   **The number line is shown in Step 2.**

Coached Example

A music video is 5 minutes long. How many seconds long is the music video?

Find the relationship between minutes and seconds.

1 minute = _____ seconds

Which is the larger unit? _____

Which is the smaller unit? _____

When you change from a larger unit to a smaller unit, which operation do you use?
_____

Show your work.

There are _____ seconds in 5 minutes.

The music video is _____ seconds long.

Lesson Practice

Choose the correct answer.

1. Alana started studying at 3:30 P.M. and finished at 6:00 P.M. How much time did Alana spend studying?

   A. 2 hours 30 minutes

   B. 3 hours

   C. 3 hours 2 minutes

   D. 3 hours 30 minutes

2. A play started at 11:10 A.M. and ended at 12:50 P.M. How long was the play?

   A. 50 minutes

   B. 1 hour 30 minutes

   C. 1 hour 40 minutes

   D. 2 hours 20 minutes

3. Mark left his house at 1:05 P.M. He returned to his house at the time shown on the clock below.

   How long was Mark away from home?

   A. 4 hours 18 minutes

   B. 3 hours 18 minutes

   C. 3 hours 8 minutes

   D. 2 hours 8 minutes

4. At 7:15 P.M., Will finished watching a movie that lasted 1 hour 30 minutes. At what time did the movie start?

   A. 5:45 P.M.

   B. 6:45 P.M.

   C. 8:15 P.M.

   D. 8:45 P.M.

5. Samantha has 10 minutes to finish a puzzle. How many seconds are in 10 minutes?

   A. 60 seconds

   B. 100 seconds

   C. 160 seconds

   D. 600 seconds

6. When elected, a United States senator serves a 6-year term. How many months does a senator serve?

   A. 84 months

   B. 72 months

   C. 60 months

   D. 36 months

7. Peter spent 8 days on his vacation. He spent $\frac{1}{2}$ of his vacation at the beach. How many days of his vacation did he spend at the beach?

   A. 2 days

   B. 3 days

   C. 4 days

   D. 6 days

8. Sophie's road trip from Pennsylvania to Maryland took a total of 6 hours. She spent $\frac{1}{3}$ of the time at an outlet mall along the way.

   A. How much time, in hours, did Sophie spend at the outlet mall? Show your work. Then show the amount of time on the number line.

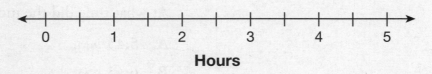

   B. How much time, in minutes, did Sophie spend at the outlet mall? Show your work.

Weight and Mass

Common Core State Standards:
4.MD.1, 4.MD.2

Getting the Idea

Weight is the measure of how heavy an object is. You can measure weight using a scale or a balance. Units of weight are in the customary system.

You can use these benchmarks to estimate weight.

A slice of bread weighs about 1 ounce.

A loaf of bread weighs about 1 pound.

Example 1

Kenika has a cell phone in her pocket. Does her cell phone weigh about 8 pounds or about 8 ounces?

Strategy   **Use benchmarks.**

Step 1   Think about something that weighs about 1 pound.
          A loaf of bread weighs 1 pound.

Step 2   Does a cell phone weigh as much as 8 loaves of bread?
          No, a cell phone weighs less than 8 pounds.

Step 3   Think about something that weighs about 1 ounce.
          A slice of bread weighs about 1 ounce.

Step 4   Does a cell phone weigh as much as 8 slices of bread?
          Yes, a cell phone weighs about 8 ounces.

Solution   **Kenika's cell phone weighs about 8 ounces.**

When you solve a word problem about weight, write a number sentence to represent the problem.

Example 2

Alex bought 2 bags of potatoes. Each bag weighs $2\frac{1}{2}$ pounds.

How many pounds of potatoes in all did Alex buy?

Strategy   **Write a number sentence.**

Step 1   Write a number sentence to represent the problem.

He bought 2 bags of potatoes that weight $2\frac{1}{2}$ pounds each.

Let p represent the total weight of the potatoes.

$2\frac{1}{2} + 2\frac{1}{2} = p$

Step 2   Add the mixed numbers.

Add the fraction parts first.
Then add the whole number parts.

$$\begin{array}{r} 2\frac{1}{2} \\ + 2\frac{1}{2} \\ \hline 4\frac{2}{2} \end{array}$$

Step 3   Simplify mixed number $4\frac{2}{2}$.

Simplify $\frac{2}{2} = 1$.

Add to the whole number, $4 + 1 = 5$.

Solution   **Alex bought 5 pounds of potatoes in all.**

You can show a weight on a number line.

Example 3

Terry ordered $1\frac{1}{4}$ pounds of roast beef at the deli counter.

Show the amount of roast beef on a number line.

Strategy   **Label a number line in fourths.**

Step 1   Draw a number line from 0 to 2. Label it in fourths.
Label the units.

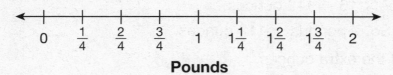

Pounds

Step 2   Label the amount of roast beef that Terry bought.
She bought $1\frac{1}{4}$ pounds of roast beef.
So label a point at $1\frac{1}{4}$ on the number line.

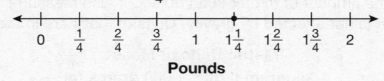

Pounds

Solution   **The number line is shown in Step 2.**

The table shows the relationship between pounds and ounces.

Customary Units of Weight
1 **pound (lb)** = 16 **ounces (oz)**

When you change from a larger unit to a smaller unit, use multiplication or addition.

When you change from a smaller unit to a larger unit, use division or subtraction.

Example 4

Drew weighed 7 pounds 9 ounces when he was born.

How many ounces are in 7 pounds 9 ounces?

Strategy   **Use multiplication to change from pounds to ounces. Then add the extra ounces.**

| Step 1 | Find the relationship between pound and ounces. |

1 pound = 16 ounces

| Step 2 | Multiply 7 pounds by 16 ounces. |

$7 \times 16 = 112$ ounces

So, 7 pounds = 112 ounces.

| Step 3 | Add the extra ounces. |

$112 + 9 = 121$ ounces

Solution   **There are 121 ounces in 7 pounds 9 ounces.**

Mass measures the amount of matter in an object. It also measures how heavy an object is, except it is not affected by gravity. Units of mass are in the metric system.

Metric Units of Mass
1 **kilogram (kg)** = 1,000 **grams (g)**

You can use these benchmarks to estimate mass.

A pen cap has a mass of about 1 gram.

A textbook has a mass of about 1 kilogram.

Example 5

The table shows the relationship between the number of kilograms and the number of grams. Complete the rest of the table.

Kilograms	Grams
1	1,000
2	2,000
3	
4	
5	

Strategy   **Use the relationship between kilograms and grams.**

Step 1   Find the relationship between kilograms and grams.

  1 kilogram = 1,000 grams

Step 2   Multiply 2 kilograms by 1,000 grams.

  2 × 1,000 = 2,000 grams

  So, 2 kilograms = 2,000 grams.

Step 3   Find the number of grams in 3 to 5 kilograms.

  Multiply the number of grams by 1,000 grams.

  3 × 1,000 = 3,000 grams

  4 × 1,000 = 4,000 grams

  5 × 1,000 = 5,000 grams

Step 4   Complete the table.

Kilograms	Grams
1	1,000
2	2,000
3	3,000
4	4,000
5	5,000

Solution   **The table is shown in Step 4.**

Example 6

Jake caught a fish with a mass of 2 kilograms. Matthew caught a fish with a mass of 1,750 grams. Whose fish has a greater mass?

Strategy   **Change 2 kilograms to grams. Then compare.**

Step 1   Multiply to change 2 kilograms to grams.

2 × 1,000 grams = 2,000 grams

Step 2   Compare the masses.

Jake's fish: 2 kilograms or 2,000 grams

Matthew's fish: 1,750 grams

2,000 grams > 1,750 grams

Solution   **Jake's fish has a greater mass.**

Coached Example

Nicole has 3 pounds of peanuts and 45 ounces of raisins. Does she have more peanuts or raisins?

Which is the smaller unit, pounds or ounces? _____

Find the relationship between pound and ounces.

1 pound = _____ ounces

Multiply to change 3 pounds to ounces.

3 × _____ = _____ ounces

Compare the weights.

Peanuts: 3 pounds or _____ ounces

Raisins: 45 ounces

_____ ounces ◯ 45 ounces

Nicole has more _____ than _____.

Lesson Practice

Choose the correct answer.

1. Which could be the weight of a desk?

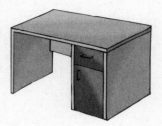

  A. 4 ounces   **C.** 40 ounces

  B. 4 pounds   **D.** 40 pounds

2. Which is the best estimate for the mass of a cat?

  A. 30 kilograms

  B. 3 kilograms

  C. 30 grams

  D. 3 grams

3. A brick has a mass of 3 kilograms. A rock has a mass of 2,500 grams. Which sentence is true?

  A. The rock has more mass than the brick.

  B. The brick has less mass than the rock.

  C. The rock has less mass than the brick.

  D. The brick has the same amount of mass as the rock.

4. A chair has a mass of 7 kilograms. What is the mass, in grams, of the chair?

  A.   70 grams

  B.   112 grams

  C.   700 grams

  D. 7,000 grams

5. Which table shows the relationship between pounds and ounces?

A.

Pounds	Ounces
2	16
3	32
4	64

B.

Pounds	Ounces
2	32
3	48
4	64

C.

Pounds	Ounces
2	32
3	64
4	96

D.

Pounds	Ounces
2	20
3	30
4	40

6. Ruben buys a 3-pound 4-ounce bag of apples. How many ounces are there in 3 pounds 4 ounces?

A. 52 ounces

B. 48 ounces

C. 44 ounces

D. 34 ounces

7. The mass of a nickel is 5 grams. What is the mass, in grams, of 40 nickels?

A. 20 grams

B. 40 grams

C. 200 grams

D. 400 grams

8. A box of cards has 12 birthday cards. Each card weighs 2 ounces. How much do the 12 cards weigh?

A. 1 pound 4 ounces

B. 1 pound 8 ounces

C. 2 pound 4 ounces

D. 2 pound 8 ounces

9. Chelsea mailed 2 boxes and 2 bags to her uncle in the army. The two boxes weighed $8\frac{1}{2}$ pounds each and the two bags weighed $4\frac{1}{2}$ pounds each.

A. How many pounds do all 2 boxes and 2 bags weigh? Show your work.

_____

_____

B. What is the total weight, in ounces, of the two bags? Show your work.

_____

_____

Capacity

Common Core State Standards:
4.MD.1, 4.MD.2

Getting the Idea

Capacity or **liquid volume** measures how much liquid a container holds.
You can use these benchmarks to estimate capacity.

Customary Units of Capacity

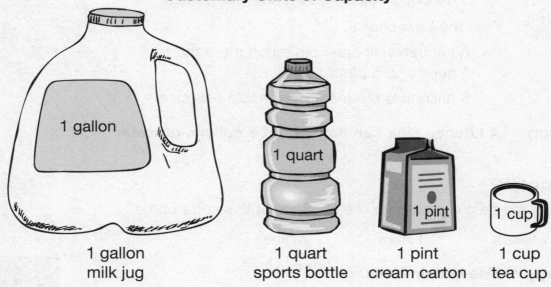

| 1 gallon
milk jug | 1 quart
sports bottle | 1 pint
cream carton | 1 cup
tea cup |

Metric Units of Capacity

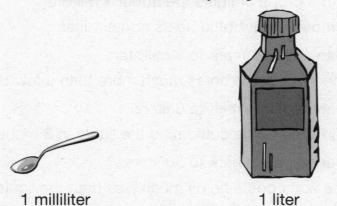

1 milliliter
of water in a teaspoon

1 liter
mouthwash bottle

Example 1

Which is the best estimate for the amount of water a kitchen sink can hold?

5 gallons   5 pints   5 cups

Strategy   **Compare an actual sink to 1 gallon.**

Step 1   Think about an actual kitchen sink.

A kitchen sink is big and can hold at least 1 gallon.

Step 2   Pick the best choice.

A kitchen sink can hold much more than 5 cups and 5 pints.

5 gallons is the most reasonable estimate.

Solution   **A kitchen sink can hold about 5 gallons of water.**

Example 2

Which is the best estimate for the capacity of this soda bottle?

3 milliliters   3 liters   30 liters

Strategy   **Use benchmarks.**

Step 1   Think about benchmarks for capacity.

3 to 4 drops of liquid are about 1 milliliter.

A mouthwash bottle holds about 1 liter.

Step 2   Compare a benchmark to 3 milliliters.

The soda bottle holds much more than a few drops of liquid.

Step 3   Compare a benchmark to 3 liters.

The soda bottle could hold the liquid in 3 mouthwash bottles.

Step 4   Compare a benchmark to 30 liters.

The soda bottle holds much less than the liquid in 30 mouthwash bottles.

Solution   **The best estimate for the capacity of the soda bottle is 3 liters.**

The table shows the relationships of the customary units of capacity.

Customary Units of Capacity
1 **pint (pt)** = 2 **cups (c)**
1 **quart (qt)** = 2 pints
1 **gallon (gal)** = 4 quarts

When you change from a larger unit to a smaller unit, use multiplication or addition. When you change from a smaller unit to a larger unit, use division or subtraction.

Example 3

Jason has 8 quarts of water at home. He buys 2 gallons of water at the store. How many gallons of water does Jason have now?

Strategy   **Use division to change from quarts to gallons. Then add the gallons.**

Step 1   Find the relationship between quarts and gallons.

4 quarts = 1 gallon

Step 2   Divide 8 quarts by 4 quarts.

8 ÷ 4 = 2 gallons

So, Jason bought 8 quarts or 2 gallons of water at the store.

Step 3   Add the gallons.

2 + 2 = 4 gallons

Solution   **Jason has 4 gallons of water now.**

The table shows the relationships of the customary units of capacity.

Metric Units of Capacity
1 **liter (L)** = 1,000 **milliliters (mL)**

Example 4

The table shows the relationship between the number of liters and the number of milliliters. Complete the table.

Liters	Milliliters
2	
4	
6	
8	

Strategy   **Use the relationship between liters and milliliters.**

Step 1   Find the relationship between liters and milliliters.

1 liter = 1,000 milliliters

Step 2   How many milliliters are in 2 liters?

Multiply 2 liters by 1,000 milliliters.

2 × 1,000 = 2,000 milliliters

Step 3   Find the number of milliliters in 4, 6, and 8 liters.

Multiply the number of liters by 1,000 milliliters.

4 × 1,000 = 4,000 milliliters

6 × 1,000 = 6,000 milliliters

8 × 1,000 = 8,000 milliliters

Step 4   Complete the table.

Liters	Milliliters
2	2,000
4	4,000
6	6,000
8	8,000

Solution   **The table is shown in Step 4.**

Example 5

Mrs. O'Brien poured 3 liters of paint into 5 canisters. She poured the same amount of paint into each canister. About how much paint, in milliliters, is in each canister?

Strategy   **Multiply to change liters to milliliters. Then divide.**

| Step 1 | Use multiplication to change 3 liters to milliliters. |

1 liter = 1,000 milliliters

$3 \times 1,000 = 3,000$ milliliters

So 3 liters = 3,000 milliliters.

| Step 2 | Find the amount of paint in each canister. |

There is a total of 3,000 milliliters. There are 5 canisters.

Let c represent the amount of paint in each canister.

Find $3,000 \div 5 = c$.

$3,000 \div 5 = 600$ milliliters

Solution   **Each canister has about 600 milliliters of paint.**

Coached Example

A bottle has a capacity of 2 liters. A bucket has a capacity that is 4 times more than the bottle. What is the capacity, in milliliters, of the bucket?

Write a number sentence for the problem.

The bottle has a capacity of _____ liters.

The bucket has a capacity that is _____ times more than the bottle.

Which operation should you use to find the capacity of the bucket? _____

Let b represent the capacity of the bucket.

Find _____ × _____ = b

Multiply.

_____ × _____ = _____ liters

Change the capacity of the bucket in liters to milliliters.

1 liter = _____ milliliters

Multiply to change from liters to milliliters.

_____ × _____ = _____ milliliters

The capacity of the bucket is _____ milliliters.

Lesson Practice

Choose the correct answer.

1. Which is the best estimate for the amount of orange juice in the glass?

   A. 50 liters   **C.** 250 liters

   B. 50 milliliters   **D.** 250 milliliters

2. Which object holds about 1 liter of water?

   A.

   B.

   C.

   D.

3. Which container's capacity would be best to measure in milliliters?

   A. bathtub

   B. coffee cup

   C. swimming pool

   D. water cooler

4. Michael bought 7 quarts of engine oil for his truck. How many pints are in 7 quarts?

   A. 12 pints

   B. 14 pints

   C. 28 pints

   D. 56 pints

5. Kate fills a tank with 5 liters of water. How many milliliters are in 5 liters?

   A. 5 milliliters

   B. 50 milliliters

   C. 500 milliliters

   D. 5,000 milliliters

6. A glass has a capacity of 750 milliliters. A pitcher has a capacity that is 5 times more than the capacity of the glass. What is the capacity of the pitcher?

   A. 3,750 milliliters

   B. 5,000 milliliters

   C. 5,750 milliliters

   D. 7,500 milliliters

7. Maria drank 6 cups of water yesterday and 8 cups of water today. How many pints of water did Maria drink in all?

   A. 2 pints

   B. 6 pints

   C. 7 pints

   D. 8 pints

8. Carol poured a total of 4 liters of iced tea into 8 tumblers. Each tumbler has the same amount of iced tea. How much iced tea, in milliliters, is in each tumbler?

   A. 200 milliliters

   B. 400 milliliters

   C. 450 milliliters

   D. 500 milliliters

9. Eliot has a 3-liter punch bowl. He poured 850 mL of pineapple juice, 900 mL of orange juice, and 250 mL of ginger ale into the bowl.

   A. What is the capacity of the punch bowl in milliliters? Show your work.

   _____

   B. How much liquid, in liters, did Eliot pour into the bowl in all? Show your work.

   _____

Common Core State Standards:
4.MD.1, 4.MD.2

Length

Getting the Idea

Length measures how long, wide, or tall an object is. It also measures distances.
The table below shows some units of length in the customary system.

Customary Units of Length
1 **foot (ft)** = 12 **inches (in.)**
1 **yard (yd)** = 3 feet
1 **mile (mi)** = 1,760 yards

You can use these benchmarks to estimate lengths.

A 12-inch ruler measures 1 foot.

A yardstick measures 1 yard.

An adult can walk 1 mile in about 20 minutes.

Example 1

Which real object is most likely to be 150 feet long?

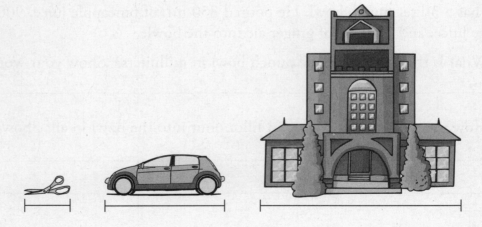

Strategy **Think about the length of 1 foot.**

Step 1 Think about how long 150 feet will be.

The length of this book is about 1 foot.

Imagine lining up 150 books to get 150 feet.

Step 2   Review the choices.

> A pair of scissors is much shorter than 150 feet.
>
> A 4-door car is shorter than 150 feet.
>
> A building could be 150 feet long.

Solution   **The building is most likely about 150 feet.**

The table below shows some units of length in the metric system.

Metric Units of Length
1 **centimeter (cm)** = 10 **millimeters (mm)**
1 **meter (m)** = 100 centimeters
1 **kilometer (km)** = 1,000 meters

This line measures 1 centimeter. ____

1 meter is a little shorter than 1 yard.

An adult can walk 1 kilometer in about 10 minutes.

Example 2

Which real object is most likely to be 20 milliliters long?

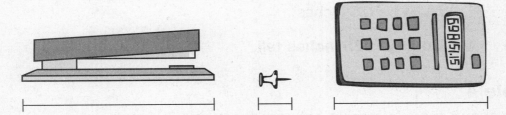

Strategy   **Think about the length of 1 millimeter.**

Step 1   Think about how long 20 millimeters will be.

> I millimeter is about the thickness of a dime.
>
> 20 millimeters will be about the height of 20 stacked dimes.

Step 2   Review the choices.

> A stapler is longer than 20 millimeters.
>
> The pushpin could be about 20 millimeters.
>
> A calculator is about the length of a stapler, so it is longer
> than 20 millimeters.

Solution   **The pushpin is most likely about 20 millimeters.**

You can use the relationship between units to change from one unit to another.

When you change a larger unit to a smaller unit, use multiplication or addition.
To change 3 feet to inches, multiply 3 × 12.
So 3 feet = 36 inches.

When you change a smaller unit to a larger unit, use division or subtraction.
To change 100 millimeters to centimeters, divide 100 ÷ 10.
So 100 millimeters = 10 centimeters.

Example 3

Mr. Conroy is 6 feet and 3 inches tall. How tall is Mr. Conroy in inches?

Strategy   **Multiply to change feet to inches. Then add the extra inches.**

Step 1   Find the relationship between feet and inches.
              1 foot = 12 inches

Step 2   Multiply 6 feet by 12 inches.
              6 × 12 = 72 inches

Step 3   Add the extra inches.
              72 + 3 = 75 inches

Solution   **Mr. Conroy is 75 inches tall.**

Example 4

The table shows the relationship between the number of meters and the number of centimeters. Complete the table.

Meters	Centimeters
2	
4	
6	
8	

Strategy   **Use the relationship between meters and centimeters.**

Step 1   Find the relationship between meters and centimeters.
              1 meter = 100 centimeters

Step 2 How many centimeters are in 2 meters?

Multiply 2 meters by 100 centimeters.

$2 \times 100 = 200$ centimeters.

So 2 meters $= 200$ centimeters.

Step 3 Find the number of centimeters in 4, 6, and 8 meters.

$4 \times 100 = 400$ centimeters

$6 \times 100 = 600$ centimeters

$8 \times 100 = 800$ centimeters

Step 4 Complete the table.

Meters	Centimeters
2	200
4	400
6	600
8	800

Solution **The table is shown in Step 4.**

You can use a number line to represent length.

Example 5

Troy jogged $1\frac{2}{3}$ miles this morning.

Show the distance Troy jogged on a number line.

Strategy **Make equal parts of fractional lengths on a number line.**

Step 1 Draw a number line from 0 to 2.

The denominator is 3, so draw the number line in thirds.

Step 2 Find $1\frac{2}{3}$ on the number line. Draw a point.

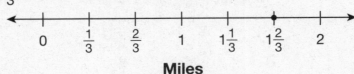

Miles

Solution **The number line is shown in Step 2.**

When you solve a real world problem, write a number sentence to represent the problem.

Example 6

A window curtain is $3\frac{3}{8}$ feet wide. What is the total width of two window curtains side by side?

Strategy **Write a number sentence for the problem.**

Step 1 Write a number sentence.

Each curtain is $3\frac{3}{8}$ feet wide.

To find the total width, use addition.

Let w represent the total width of two curtains.

Find $3\frac{3}{8} + 3\frac{3}{8} = w$.

Step 2 Find the sum.

Add the fraction parts.
Then add the whole number parts.

$$3\frac{3}{8}$$
$$+\ 3\frac{3}{8}$$
$$\overline{\quad 6\frac{6}{8}\quad}$$

Step 3 Simplify.

$$\frac{6}{8} = \frac{6 \div 2}{8 \div 2} = \frac{3}{4}$$

So $6\frac{6}{8} = 6\frac{3}{4}$.

Solution **The total width of two curtains is $6\frac{3}{4}$ feet.**

Coached Example

Nicole lives 3 kilometers from the mall and 1.6 kilometers from her school. How far, in meters, does Nicole live from the mall?

On the number line below, show the distance Nicole lives from her school.

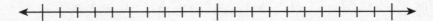

Kilometers

Find how far Nicole lives from the mall.

She lives _____ kilometers from the mall.

To change from kilometers to meters, should you use multiplication or division?

_____

1 kilometer = _____ meters

3 × _____ meters = _____ meters

So 3 kilometers = _____ meters.

Show the distance Nicole lives from school on the number line.

She lives _____ kilometers from the school.

The number line is in tenths. Label 0, 1, and 2 on the number line.

Find 1.6 on the number line. Draw a point.

Nicole lives _____ meters from the mall. The point on the number line above shows the distance, in kilometers, Nicole lives from school.

Lesson Practice

Choose the correct answer.

1. Which measure is most likely the height of a ceiling?

   A. 10 miles

   B. 10 yards

   C. 10 feet

   D. 10 inches

2. Which measure is most likely the length of a digital camera?

   A. 12 millimeters

   B. 12 centimeters

   C. 12 meters

   D. 12 kilometers

3. Which measure is equal to 1,000 centimeters?

   A. 10 meters

   B. 10 millimeters

   C. 100 meters

   D. 100 millimeters

4. A lamp is 4 feet 8 inches tall. What is the height of the lamp in inches?

   A. 40 inches

   B. 48 inches

   C. 52 inches

   D. 56 inches

5. The point on the number line below shows the length of a piece of chalk.

   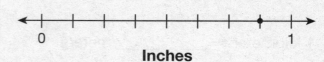

   Inches

   What is the length of the chalk?

   A. $\frac{1}{8}$ inch   **C.** $\frac{7}{8}$ inch

   B. $\frac{1}{4}$ inch   **D.** $\frac{9}{10}$ inch

6. The point on the number line below shows height of a ladder.

   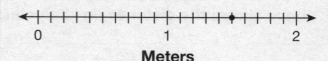

   Meters

   What is the height of the ladder?

   A. 1.5 meters

   B. 1.6 meters

   C. 2.1 meters

   D. 2.5 meters

7. A painting is 4 feet long. A photo frame is 10 inches long. How many inches longer is the painting than the photo frame?

    A.   6 inches

    B. 24 inches

    C. 30 inches

    D. 38 inches

8. LeAnne placed a box that is $1\frac{3}{4}$ feet high on top of another box that is $1\frac{1}{4}$ feet high. How many feet high are the boxes when stacked?

    A. $2\frac{3}{4}$ feet

    B. 3 feet

    C. $3\frac{1}{2}$ feet

    D. 4 feet

9. Daron has a cord that is 12 inches long and another one that is $2\frac{1}{2}$ feet. He connects the two cords.

    A. What is the total length, in inches, of the two cords? Show your work.

    _____

    _____

    _____

    B. In the number line below, show the total length, in feet, of the two cords. Explain how you found the total length in feet.

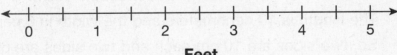

Feet

    _____

    _____

Perimeter

Common Core State Standard:
4.MD.3

Getting the Idea

Perimeter is the measure of the distance around a figure. The perimeter is measured in customary or metric units of length, such as inches, feet, centimeters, or meters.

To find the perimeter of a figure, add the lengths of all the sides.

Example 1

What is the perimeter of this rectangle?

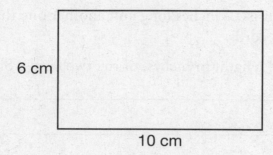

6 cm

10 cm

Strategy   **Add the lengths of the sides.**

> **Step 1**   Find all the side lengths of the rectangle.
>
> A rectangle has 4 sides, with opposite sides having equal lengths.
>
> The length is 10 centimeters and the width is 6 centimeters.
>
> So, two sides are 10 cm each and two sides are 6 cm each.

> **Step 2**   Add the lengths of the four sides.
>
> 10 cm + 10 cm + 6 cm + 6 cm = 32 cm

Solution   **The perimeter of the rectangle is 32 centimeters.**

For Example 1, you could also use the formula for the perimeter of a rectangle.

Perimeter = (2 × length) + (2 × width)

$P = (2 \times 10) + (2 \times 6)$

$P = 20 + 12$

$P = 32$ cm

You can use a variable, such as *l* or *w*, to represent the length or the width of a rectangle.

Example 2

The rectangle below has a perimeter of 30 inches.

8 inches

P = 30 inches

What is the width of the rectangle?

Strategy   **Use the formula for the perimeter of a rectangle.**

Step 1   Substitute the values into the formula for perimeter.

The perimeter is 30 feet.

The length is 8 feet.

Use *w* to represent the width.

Perimeter = (2 × length) + (2 × width)

$30 = (2 \times 8) + (2 \times w)$

Step 2   Multiply the values inside the parentheses.

$30 = (2 \times 8) + (2 \times w)$

$30 = 16 + 2 \times w$

Step 3   Subtract 16 on both sides of the equal sign.

$30 = 16 + 2 \times w$

$30 - 16 = 16 - 16 + 2 \times w$

$14 = 2 \times w$

Step 4   Divide both sides of the equal sign by 2 to solve for *w*.

$14 = 2 \times w$

$14 \div 2 = 2 \div 2 \times w$

$7 = 1 \times w$

$7 = w$

Solution   **The width of the rectangle is 7 inches.**

For Example 2, you can check the answer by substituting 7 inches for the width in the perimeter formula.

Perimeter = (2 × length) + (2 × width)

$$P = (2 \times 8) + (2 \times 7)$$

$$P = 16 + 14$$

$$P = 30 \text{ in.}$$

The perimeter is 30 inches. So, the answer is correct.

A square is a rectangle with all 4 sides equal in length. To find the perimeter of a square, add the lengths of all the sides. You could also multiply the length of one side by 4.

Here is the formula for the perimeter of a square:

Perimeter = 4 × length of a side

$$P = 4 \times s$$

Example 3

What is the perimeter of this square?

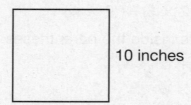

10 inches

Strategy   **Use the formula for the perimeter of a square.**

$$P = 4 \times s$$

$$P = 4 \times 10$$

$$P = 40 \text{ inches}$$

Solution   **The perimeter of the square is 40 inches.**

Example 4

A balcony has a perimeter of 28 meters. The balcony is in the shape of a square. What is the length of one side of the balcony?

Strategy   **Use the formula for the perimeter of a square.**

Step 1   The balcony is a square. Write the formula for the perimeter.

$$P = 4 \times s$$

Step 2   Substitute the values you know into the formula.

The perimeter is 28 meters.

Use the variable s to represent the side length.

$$P = 4 \times s$$

$$28 = 4 \times s$$

Step 3   Divide both sides of the equal sign by 4 to solve for s.

$$28 = 4 \times s$$

$$28 \div 4 = 4 \div 4 \times s$$

$$7 = 1 \times s$$

$$7 = s$$

Solution   **The length of one side of the balcony is 7 meters.**

Coached Example

The rectangle below has a perimeter of 60 inches and a width of 10 inches.

P = 60 inches	10 inches

What is the length of the rectangle?

Write the formula for the perimeter of a rectangle.

$$P = (2 \times \underline{\hspace{4cm}}) + (2 \times \underline{\hspace{4cm}})$$

Substitute the values you know into the formula.

Use the variable *l* to represent the _____.

$$60 = (2 \times \underline{\hspace{2cm}}) + (2 \times \underline{\hspace{2cm}})$$

Solve for the length.

Check your answer by substituting _____ inches for the length in the formula.

$$P = (2 \times \underline{\hspace{2cm}}) + (2 \times \underline{\hspace{2cm}})$$

$$P = \underline{\hspace{2cm}} + \underline{\hspace{2cm}}$$

$$P = \underline{\hspace{2cm}}$$

The length of the rectangle is _____ inches.

Lesson Practice

Choose the correct answer.

1. What is the perimeter of this square?

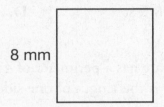

8 mm

   A. 16 mm
   B. 32 mm
   C. 48 mm
   D. 64 mm

2. What is the perimeter of this rectangle?

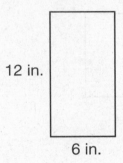

12 in.

6 in.

   A. 18 in.
   B. 30 in.
   C. 36 in.
   D. 72 in.

3. The square below has a perimeter of 36 meters.

P = 36 m

   What is the length of one side of the square?

   A. 4 m
   B. 8 m
   C. 9 m
   D. 12 m

4. The rectangle below has a perimeter of 72 centimeters and a length of 21 centimeters.

21 cm

P = 72 cm

   What is the width of the rectangle?

   A. 15 cm
   B. 30 cm
   C. 42 cm
   D. 51 cm

5. A rectangular playground has a length of 75 yards and width of 50 yards. What is the perimeter of the playground?

    A. 125 yards   **C.** 200 yards

    B. 175 yards   **D.** 250 yards

6. The floor of Jimmy's tree house is in the shape of a square. Each side of the floor is 15 feet. What is the perimeter of Jimmy's tree house?

    A. 30 feet   **C.** 90 feet

    B. 60 feet   **D.** 225 feet

7. Richie has a rug in his bedroom with a perimeter of 42 feet. The length of the rug is 12 feet. What is the width of the rug?

    A. 30 feet   **C.** 9 feet

    B. 18 feet   **D.** 6 feet

8. A square has a perimeter of 44 meters. What is the length of one side of the square?

    A. 9 meters   **C.** 11 meters

    B. 10 meters   **D.** 12 meters

9. Anna drew the square and the rectangle below.

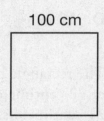

100 cm   13 km

P = 46 km

    A. What is the perimeter of the square? Show your work.

    B. What is the width of the rectangle? Show your work.

Common Core State Standard:
4.MD.3

Area

Getting the Idea

Area is the measure of the region inside a figure. Area is measured in square units, such as square inches, square feet, and square centimeters. A square inch, for example, is a square with a side length of 1 inch.

To find the area of a figure, count the number of square units inside the figure. A scale tells what each square unit represents.

Example 1

What is the area of this rectangle?

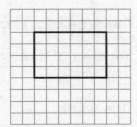

Scale: □ = 1 square meter

Strategy   **Count the number of square units inside the figure.**

Step 1   Count the square units.

There are 24 square units inside the rectangle.

Step 2   Look at the scale to find what each square unit represents.

Each square unit is 1 square meter.

So, 24 square units = 24 square meters.

Solution   **The area of the rectangle is 24 square meters.**

You can use formulas to find the areas of rectangles and squares.

To find the area of a rectangle, multiply the length by the width.

Area = length × width

$$A = l \times w$$

Example 2

What is the area of this rectangle?

8 cm

5 cm

Strategy   **Use the formula for the area of a rectangle.**

Step 1   Write the formula for the area of a rectangle.

$$A = l \times w$$

Step 2   Multiply the length by the width.

The length is 5 centimeters.

The width is 8 centimeters.

$A = 5 \times 8 = 40$ square centimeters

Solution   **The area of the rectangle is 40 square centimeters.**

To find the area of a square, multiply the length of one side by itself.

Area = side × side

$$A = s \times s$$

Example 3

What is the area of this square?

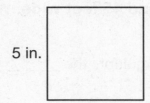

5 in.

Strategy   **Use the formula for the area of a square.**

Step 1   Write the formula for the area of a square.

$A = s \times s$

Step 2   Multiply the length of one side by itself.

The length is 5 inches.

$A = 5 \times 5 = 25$ square inches

Solution   **The area of the square is 25 square inches.**

Example 4

The floor of Winnie's bedroom is a rectangle. It has an area of 108 square feet. The length is 9 feet. What is the width of Winnie's bedroom?

Strategy   **Use the formula for the area of a rectangle.**

Step 1   The bedroom floor is a rectangle, so write the formula for the area.

$A = l \times w$

Step 2   Substitute the values into the formula.

The area is 108 square feet.

The length is 9 feet.

$108 = 9 \times w$

Step 3   Divide both sides of the equal sign by 9 to solve for w.

$108 = 9 \times w$

$108 \div 9 = 9 \div 9 \times w$

$12 = 1 \times w$

$12 = w$

Solution   **The width of Winnie's bedroom is 12 feet.**

Coached Example

A playground is 80 feet long and 45 feet wide. What is the area of the playground?

To find the area of the rectangle multiply the _____ by the _____.

Write the area formula. Use *l* for length and *w* for width.

$A = $ _____ $\times$ _____

Substitute the values into the formula.

$A = $ _____ $\times$ _____

Multiply.

$A = $ _____

Label the product with the correct units.

The units are _____ _____.

The playground has an area of _____ square feet.

Lesson Practice

Choose the correct answer.

1. What is the area of this rectangle?

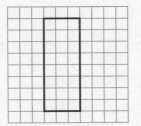

Scale: ☐ = 1 square inch

 A. 24 square inches

 B. 22 square inches

 C. 12 square inches

 D. 11 square inches

2. What is the area of this rectangle?

8 in.

12 in.

 A. 20 square inches

 B. 40 square inches

 C. 48 square inches

 D. 96 square inches

3. Each side of a square poster measures 12 inches. What is the area of the poster?

 A. 48 square inches

 B. 96 square inches

 C. 144 square inches

 D. 288 square inches

4. A rectangular table has an area of 42 square feet. The width is 3 feet. What is the length of the table?

 A. 7 feet

 B. 14 feet

 C. 18 feet

 D. 39 feet

5. A square has an area of 36 square inches. What is the length of its sides?

 A. 6 inches

 B. 9 inches

 C. 12 inches

 D. 18 inches

6. Lisa made a drawing of her kitchen. What is the area of Lisa's kitchen?

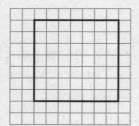

Scale: ☐ = 1 square foot

A. 14 square feet

B. 28 square feet

C. 49 square feet

D. 70 square feet

7. What is the area of the figure below?

Scale: ☐ = 1 square inch

A. 48 square inches

B. 36 square inches

C. 28 square inches

D. 24 square inches

8. A patio is in the shape of a rectangle. It has an area of 120 square feet. The length is 8 feet. What is the width of the patio?

A. 13 feet   **C.** 15 feet

B. 14 feet   **D.** 16 feet

9. Logan's office floor has an area of 700 square feet. The length is 20 feet.

A. Write an equation that can be used to find the width. Let w represent the width.

_____

B. What is the width of the office floor? Show your work.

_____

_____

_____

Angles

Common Core State Standards:
4.MD.5.a, 4.MD.5.b, 4.MD.6, 4.MD.7

Getting the Idea

An **angle** (∠) is formed by two **rays** that meet at the same **endpoint**. That endpoint is the **vertex** of the angle. An angle can be named by its vertex. The angle below can be named as angle Y or ∠Y.

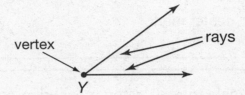

The vertex of an angle can be at the center of a circle. A **degree(°)** is the angle made by $\frac{1}{360}$ of a full turn around a circle. A full turn around a circle is 360 degrees.

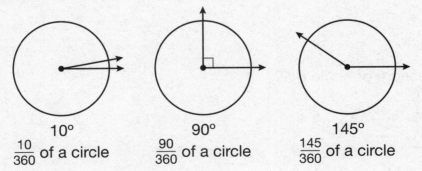

360°

The measure of an angle is the fraction of the circle between the points where the two rays intersect the circle. Below are some examples.

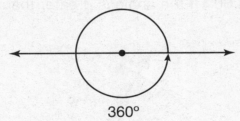

10°
$\frac{10}{360}$ of a circle

90°
$\frac{90}{360}$ of a circle

145°
$\frac{145}{360}$ of a circle

You can use a **protractor** to measure angles. A protractor often has two scales. The scales increase from 0° to 180°, but in opposite directions.

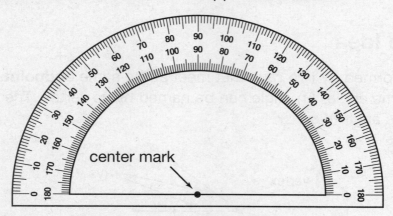

center mark

To help you decide which scale to read when measuring an angle, compare the angle to 90°.

For example, if the scales read 120° and 60°, and the angle is less than 90°, then the measure of the angle is 60°. If the angle is greater than 90°, then the measure of the angle is 120°.

Example 1
What is the measure of angle *B*?

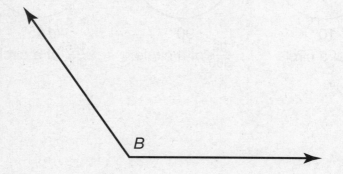

B

Strategy   **Use a protractor.**

Step 1   Place the center mark of the protractor on the vertex of the angle.

Line up one ray of the angle with the 0° mark on one of the scales.

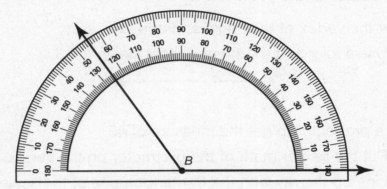

Step 2   Look at the scale at the point where the other ray of the angle crosses it.

Read the degree mark on the same scale used in Step 1.

The ray crosses the scale at 125°.

It crosses the other scale at 55°.

Step 3   Decide which scale to use.

Angle *B* appears greater than 90°, so it makes sense that the measure would be 125°, not 55°.

Solution   **The measure of angle *B* is 125°.**

Example 2

Draw an angle *S* that measures 45°.

Strategy   **Use a ruler and a protractor.**

> **Step 1**   Draw the vertex of the angle and one ray.
>
> Use a ruler to draw the ray. Label the vertex *S*.

> **Step 2**   Use a protractor to get the measure of 45°.
>
> Put the center mark of the protractor on the vertex.
>
> Line up the ray with the 0° mark on one of the scales.
>
> Then find the 45° mark. Place a dot above the protractor.

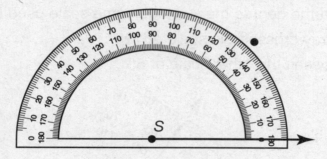

> **Step 3**   Remove the protractor. Draw the other ray.
>
> Use a ruler to draw a ray from the vertex to the dot.

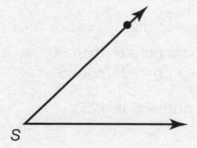

Solution   **Angle *S* is shown in Step 3.**

Example 3

The measure of angle D is 135°. A part of angle D measures 65°.

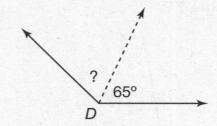

What is the measure of the other part of angle D?

Strategy   **Write a number sentence.**

Step 1   Use a variable to represent the measure of the missing part.

Choose the letter m for the missing part.

Step 2   Write a number sentence.

$$65° + m = \angle D$$
$$65° + m = 135°$$

Step 3   Solve for m.

Subtract 65° from both sides of the equal sign.

$$65° + m = 135°$$
$$65° - 65° + m = 135° - 65°$$
$$0 + m = 70°$$
$$m = 70°$$

Solution   **The measure of the other part of angle D is 70°.**

Coached Example

What is the measure of angle *T*?

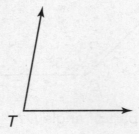

Put the center mark of the protractor on the _____ of the angle.

Line up one ray of the angle with the _____° mark on one of the scales.

Look at the scale where the other ray of the angle crosses it.

The ray crosses the scale at _____°.

It crosses the other scale at _____°.

Check your answer.

Angle *T* appears _____ than 90°, so the measure is _____°, not _____°.

The measure of angle *T* is _____°.

Lesson Practice

Choose the correct answer.

1. What is the measure of this angle?

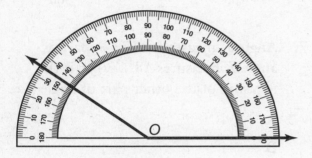

   A. 45°

   B. 55°

   C. 135°

   D. 145°

2. What is the measure of this angle?

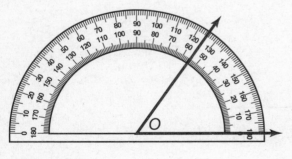

   A. 55°

   B. 60°

   C. 120°

   D. 125°

3. Which angle could measure 130°?

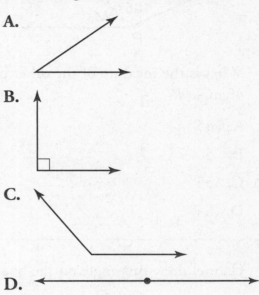

4. The measure of angle A is 90°. A part of angle A measures 35°.

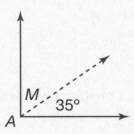

What is the measure of the other part of angle A?

   A. 65°

   B. 55°

   C. 45°

   D. 35°

5. The measure of angle *R* is 155°. A part of angle *R* measures 105°.

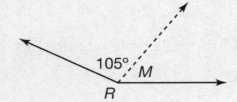

What is the measure of the other part of angle *R*?

A. 45°

B. 50°

C. 55°

D. 65°

6. A 45° angle turns through what fraction of a circle?

A. $\frac{315}{360}$

B. $\frac{45}{100}$

C. $\frac{45}{90}$

D. $\frac{45}{360}$

7. Angle *K* measures 112°. A part of angle *K* measures 48°. What is the measure of the other part of angle *K*?

A. 64°

B. 74°

C. 76°

D. 160°

8. Gabriel drew this angle on the board.

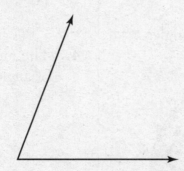

A. Use a protractor to find the measure of Gabriel's angle.

_____

B. Draw an angle that measures 45° more than Gabriel's angle.

Line Plots

Common Core State Standard:
4.MD.4

Getting the Idea

A **line plot** is a graph that uses Xs above a number line to record data. To read a line plot, count the number of Xs above the number on the number line.

Example 1

Dana and Josiah measured the lengths of the leaves they collected. They made the line plot below.

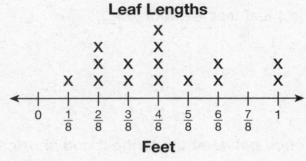

How many leaves were less than $\frac{3}{8}$ foot long?

Strategy	Count the number of Xs above each number less than $\frac{3}{8}$.

Step 1   Understand the line plot.

The number line shows the lengths, in feet.

The Xs above each number represent the number of leaves for that length.

Step 2   Count the Xs above the numbers less than $\frac{3}{8}$.

$\frac{1}{8}$ and $\frac{2}{8}$ are less than $\frac{3}{8}$.

There is 1 X above $\frac{1}{8}$ foot.

There are 3 Xs above $\frac{2}{8}$ foot.

Step 3   Add to find the total.

$1 + 3 = 4$

Solution   **There were 4 leaves that were less than $\frac{3}{8}$ foot long.**

Example 2

Use the line plot in Example 1.

What is the difference in length between the longest and the shortest leaves?

Strategy   **Find the longest and shortest lengths. Then subtract.**

Step 1   Find the length of the longest leaf.

The longest length is 1 foot.

There are 2 leaves that are 1 foot long.

Step 2   Find the length of the shortest leaf.

The shortest length is $\frac{1}{8}$ foot.

There is 1 leaf that is $\frac{1}{8}$ foot long.

Step 3   Subtract $1 - \frac{1}{8}$.

$1 = \frac{8}{8}$

$\frac{8}{8} - \frac{1}{8} = \frac{7}{8}$

Solution   **The difference between the longest and shortest leaves is $\frac{7}{8}$ foot.**

Example 3

Janelle asked some friends about the amount of orange juice they drank one day. She listed the results below.

0 cup	$\frac{1}{2}$ cup	1 cup	1 cup	0 cup	1 cup
$\frac{1}{2}$ cup	1 cup	$\frac{1}{2}$ cup	1 cup	$\frac{1}{2}$ cup	1 cup

Make a line plot of Janelle's results.

Strategy   **Count each amount. Draw a number line to make the line plot.**

Step 1 Look at the amounts of juice.

The greatest amount is 1 cup. The least amount is 0 cups.

There are also $\frac{1}{2}$ cups.

Step 2 Make a number line from 0 to 1 in halves.

Label the number line.

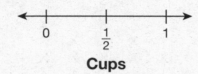

Cups

Step 3 Count the number for each amount.

2 friends drank 0 cups of juice.

4 friends drank $\frac{1}{2}$ cup of juice.

6 friends drank 1 cup of juice.

Step 4 Draw an X to represent each friend above each amount.

Write a title for the line plot.

Orange Juice Amounts

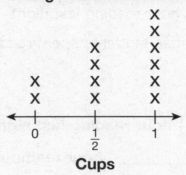

Cups

Solution **The line plot is shown in Step 4.**

Coached Example

Isaac asked some students how long they spent reading last night. He made the line plot below.

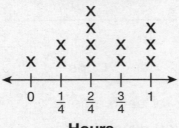

Time Spent Reading

```
                    X
                    X        X
              X     X    X   X
         X    X     X    X   X
    <----+----+-----+----+---+---->
         0    1     2    3   1
              4     4    4
```

Hours

How many students spent $\frac{1}{4}$ hour reading last night?

How much time in all did those students spend reading?

Count the number of Xs above the time of _____ hour on the number line.

There are _____ Xs above that time.

So, _____ students spent $\frac{1}{4}$ hour reading last night.

To find how much time in all those students spent reading, which operation should you use? _____

$\frac{1}{4}$ + _____ = _____

So, _____ students spent $\frac{1}{4}$ hour reading last night.

In all, those students spent _____ hour reading.

Lesson Practice

Choose the correct answer.

1. The list shows the distances that 10 students ran in gym class.

   0 miles, $\frac{1}{2}$ mile, 1 mile, 1 mile, $\frac{1}{2}$ mile,

   $\frac{1}{2}$ mile, 0 miles, 1 mile, 1 mile, $\frac{1}{2}$ mile

   Which line plot shows the results?

   A.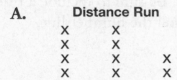
   Distance Run

   ```
         X     X
         X     X
         X     X     X
         X     X     X
      ───┼─────┼─────┼───
         0     1     1
               2
              Miles
   ```

   B. **Distance Run**

   ```
               X
         X     X     X
         X     X     X
         X     X     X
      ───┼─────┼─────┼───
         0     1     1
               2
              Miles
   ```

   C.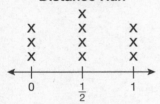
   Distance Run

   ```
               X     X
               X     X
         X     X     X
         X     X     X
      ───┼─────┼─────┼───
         0     1     1
               2
              Miles
   ```

   D.
   Distance Run

   ```
                     X
                     X
         X     X     X
         X     X     X
      ───┼─────┼─────┼───
         0     1     1
               2
              Miles
   ```

Use the line plot for questions 2–4.

William measured the lengths of some stickers. He recorded the results in a line plot.

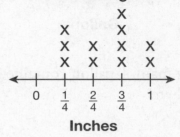

Sticker Lengths

```
                  X
         X        X
         X  X  X  X
         X  X  X  X
      ◄──┼──┼──┼──┼──►
         0  1  2  3  1
            4  4  4
           Inches
```

2. How many stickers were $\frac{3}{4}$ inch long or longer?

   A. 7   C. 5

   B. 6   D. 4

3. What is the difference in length between the longest stickers and the shortest stickers?

   A. $\frac{1}{4}$ inch   C. $\frac{3}{4}$ inch

   B. $\frac{2}{4}$ inch   D. 1 inch

4. William placed all of the $\frac{1}{4}$-inch stickers side by side. What is the total length of the stickers?

   A. $\frac{1}{4}$ inch   C. $\frac{4}{4}$ inch

   B. $\frac{3}{4}$ inch   D. $\frac{5}{4}$ inches

Use the line plot for questions 5–8.

The line plot shows the capacity of 12 containers.

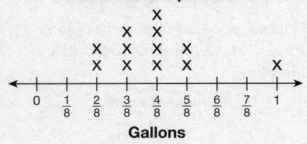

Container Capacities

Gallons

5. How many $\frac{1}{8}$-gallon containers are there?

A. 0   C. 2

B. 1   D. 3

6. How many containers are $\frac{4}{8}$ gallon or greater?

A. 2   C. 6

B. 4   D. 7

7. What is the combined capacity of all of the $\frac{2}{8}$-gallon containers?

A. $\frac{2}{8}$ gallon   C. $\frac{4}{8}$ gallon

B. $\frac{3}{8}$ gallon   D. $\frac{8}{8}$ gallon

8. What is the difference in capacity between the largest containers and the smallest containers?

A. $\frac{5}{8}$ gallon   C. $\frac{7}{8}$ gallon

B. $\frac{6}{8}$ gallon   D. $\frac{8}{8}$ gallon

9. Michelle asked some classmates how much time they spent watching a 1-hour special on TV. The list below shows the times her classmates watched the special.

$\frac{1}{4}$ hour, $\frac{2}{4}$ hour, 1 hour, $\frac{2}{4}$ hour,

$\frac{1}{4}$ hour, 1 hour, 0 hours, 1 hour,

$\frac{2}{4}$ hour, 0 hours

A. Make a line plot of Michelle's results. Be sure to include a title and label the number line.

B. How many classmates watched $\frac{2}{4}$ hour or more of the TV special?

_____

Domain 4: Cumulative Assessment for Lessons 28–36

1. The point on the number line below shows the length of a ribbon.

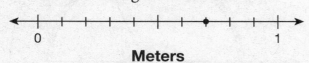

Meters

What is the length of the ribbon?

A. 0.5 meter

B. 0.6 meter

C. 0.7 meter

D. 0.8 meter

2. A rock concert lasted 2 hours and 30 minutes. How many minutes did the concert last?

A. 60 minutes

B. 90 minutes

C. 120 minutes

D. 150 minutes

3. The area of a square is 64 square meters. What is the length of one side of the square?

A. 8 meters

B. 9 meters

C. 12 meters

D. 16 meters

4. Kareem's backpack has a mass of 6 kilograms. What is the mass of the backpack in grams?

A. 60,000 grams

B. 6,000 grams

C. 600 grams

D. 60 grams

5. The length of a living room is 9 meters. How many centimeters is that?

A. 9 centimeters

B. 90 centimeters

C. 900 centimeters

D. 9,000 centimeters

6. Mitch is painting a wall in his bedroom. He knows that the area of the wall is 99 square feet. The length of the wall is 11 feet. What is the width of the wall?

A. 11 feet

B. 10 feet

C. 9 feet

D. 8 feet

7. Three children shared 4 pints of juice. How many cups are in 4 pints?

- **A.** 6 cups
- **B.** 8 cups
- **C.** 10 cups
- **D.** 16 cups

8. The angle below measures 110°. A square is placed on the angle.

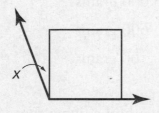

What is the measure of angle *x*?

- **A.** 20°
- **B.** 35°
- **C.** 45°
- **D.** 90°

9. What is the measure of this angle?

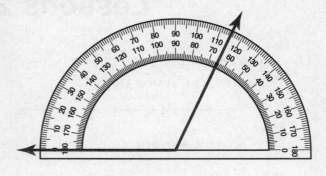

_____

10. Ron recorded the heights of some candles.

| $\frac{2}{4}$ foot | $\frac{1}{4}$ foot | $\frac{3}{4}$ foot | $\frac{3}{4}$ foot | 1 foot |
| 1 foot | $\frac{3}{4}$ foot | $\frac{2}{4}$ foot | $\frac{2}{4}$ foot | $\frac{3}{4}$ foot |

A. Create a line plot of Ron's results. Be sure to include a title and label the number line.

B. What is the difference in height between the tallest candles and the shortest candles?

_____

Domain 5 Geometry

Domain 5: Diagnostic Assessment for Lessons 37–39

1. Which shows parallel lines?

A.

B.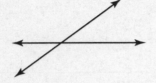

C.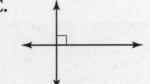

D.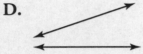

2. Which is an obtuse angle?

A.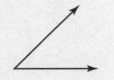

B.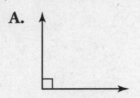

C.

D.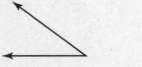

3. How many pairs of parallel sides does this figure have?

A. 3   **C.** 6

B. 4   **D.** 8

4. How many right angles does this triangle have?

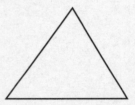

A. 0   **C.** 2

B. 1   **D.** 3

5. Jackie drew a quadrilateral.

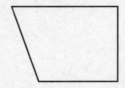

What type of quadrilateral did she draw?

A. parallelogram

B. rectangle

C. trapezoid

D. square

6. Which best describes angle *B*?

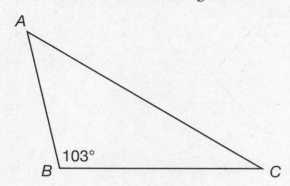

A. It is an acute angle.

B. It is a right angle.

C. It is an obtuse angle.

D. It is a straight angle.

7. Which appears to be a right triangle?

A.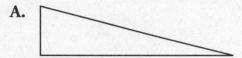

B.

C.

D.

8. Which has exactly 1 line of symmetry?

A.

B.

C.

D.

9. How many lines of symmetry does this figure have?

10. Look at the figure below.

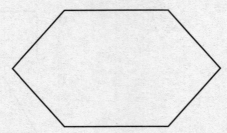

A. How many lines of symmetry does the figure have?

_____

B. Draw all the lines of symmetry on the figure below.

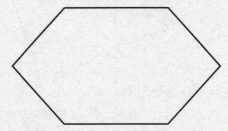

Lines and Angles

Common Core State Standard:
4.G.1

Getting the Idea

A **point** is a particular place or location.

A **line** is a straight path that goes in two directions without end.

This line with points S and T can be written as $\overleftrightarrow{ST}$ or $\overleftrightarrow{TS}$.

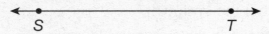

A **ray** is part of a line with one endpoint and goes in the other direction without end. A ray is named by its endpoint first. This ray is named $\overrightarrow{YZ}$.

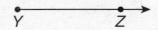

A **line segment** is part of a line with two endpoints.

This line segment can be named $\overline{MN}$ or $\overline{NM}$.

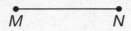

Example 1

How can you name this figure?

Strategy   **Look for endpoints or arrows.**

Step 1   Identify the figure.

The figure has arrows on both sides.

It shows a straight path without end in both directions.

The figure is a line.

Step 2   Identify points on the line.

The points B and C are on the line.

Step 3   Name the line.

Use the points on the line to name the figure.

Solution   **The figure is line *BC* or line *CB*.**

Example 2

Draw $\overrightarrow{DE}$.

Strategy   **Identify the symbol. Then use the definition to draw the figure.**

   Step 1   Identify the symbol in the name.

       The symbol above $\overrightarrow{DE}$ is ⟶.

       The figure is a ray with points D and E.

   Step 2   Draw and label the endpoint.

       The endpoint is the first point listed in the name.

       Point D is the endpoint.

   Step 3   Draw the ray.

       Draw an arrow pointing away from the endpoint.

   Step 4   Draw and label the second point on the ray.

       The other point is point E.

Solution   **Ray DE is shown in Step 4.**

Pairs of lines or line segments can be identified as parallel, intersecting, or perpendicular.

Parallel lines are lines that remain the same distance apart and never meet.

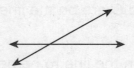

Intersecting lines are lines that cross at exactly one point.

Perpendicular lines are intersecting lines that cross to form 4 square corners.

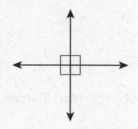

Example 3

Which street is parallel to 2nd Avenue?

Strategy	**Use the definition of parallel lines.**

Step 1 Define parallel lines.

Parallel lines are lines that remain the same distance apart.

Step 2 Find the street that is parallel to 2nd Avenue.

1st Avenue and 2nd Avenue do not intersect.

They appear to remain the same distance apart.

Solution   **1st Avenue is parallel to 2nd Avenue.**

You can draw parallel or perpendicular lines by using a ruler.

Example 4

Draw a pair of perpendicular lines.

Strategy   **Use a ruler and an object that forms a square corner.**

> **Step 1**   Use a ruler to draw a straight line.

> **Step 2**   Use an object with a square corner and place it on the line.
> You can use a book, an index card, or an envelope.

> **Step 3**   Draw a straight line that crosses the first line and forms a square corner.

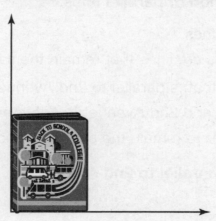

Solution   **A pair of perpendicular lines is shown in Step 3.**

Angles can be identified as acute, right, or obtuse.

A **right angle** is an angle that forms a square corner.
It measures exactly 90°.

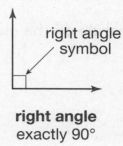

right angle
exactly 90°

An **acute angle** is an angle forms an angle less than
a right angle. It measures more than 0°, but less than 90°.

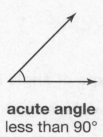

acute angle
less than 90°

An **obtuse angle** is an angle forms an angle greater than
a right angle. It measures more than 90°, but less than 180°.

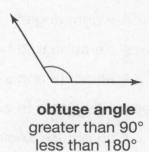

obtuse angle
greater than 90°
less than 180°

Example 5

Name the angle shown below.

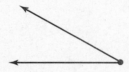

Strategy	**Compare the angle to a right angle.**
Step 1	Think about a right angle.
	A right angle measures 90° and forms a square corner.
Step 2	Compare this angle to a right angle.
	This angle is smaller than a right angle.
Solution	**The angle is an acute angle.**

Coached Example

The hands on the clock form an angle.

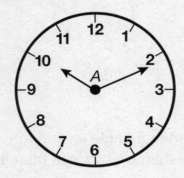

What type of angle is ∠A?

Compare the angle to a right angle.

Does ∠A appear to be exactly 90°? _____

Is ∠A a right angle? _____

Does ∠A appear to be less than 90°? _____

Is ∠A an acute angle? _____

Does ∠A appear to be greater than 90°? _____

Is ∠A an obtuse angle? _____

Angle A is a(n) _____ angle.

Lesson Practice

Choose the correct answer.

1. Which is shown below?

A. line segment *AB*

B. line *AB*

C. ray *AB*

D. ray *BA*

2. Which shows line *XY*?

A.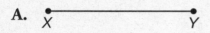

B.

C.

D.

3. Which appears to be a pair of perpendicular lines?

A.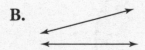

B.

C.

D.

4. Which shows ray *AB*?

A.

B.

C.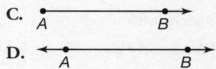

D.

5. Which is an obtuse angle?

A.

B.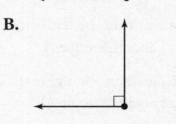

C.

D.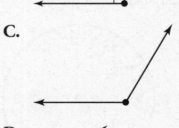

6. Which pair of lines is intersecting, but **not** perpendicular?

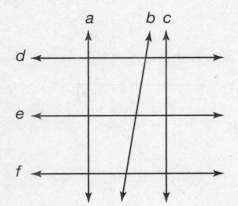

A. lines *a* and *c*

B. lines *d* and *e*

C. lines *a* and *f*

D. lines *b* and *f*

7. Which type of angle is shown below?

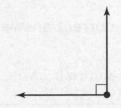

A. acute

B. right

C. obtuse

D. parallel

8. Think about the hands of an analog clock.

A. Name a time when the minute hand and hour hand form a right angle.

_____

B. In the clock on the left below, show a time where the hands would form an acute angle. Label the angle *A*.

In the clock on the right, show a time where the hands would form an obtuse angle. Label the angle *O*.

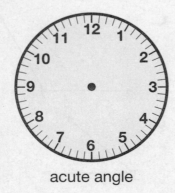

acute angle

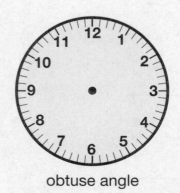

obtuse angle

Two-Dimensional Shapes

Common Core State Standards:
4.G.1, 4.G.2

Getting the Idea

A **two-dimensional shape** is a flat figure. A **polygon** is a two-dimensional shape with straight sides. Polygons are classified, or sorted, by the number of sides and angles. Each side of a polygon is a line segment. The line segments meet at points that form the angles of the polygon.

triangle	quadrilateral	pentagon	hexagon	octagon
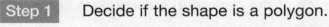				
3 sides	4 sides	5 sides	6 sides	8 sides
3 angles	4 angles	5 angles	6 angles	8 angles

A **circle** is a two-dimensional shape in which all points are an equal distance from the center. A circle is not a polygon because it does not have straight sides.

Example 1

What is the name of this two-dimensional shape?

Strategy   **Count the number of sides and angles.**

  Step 1   Decide if the shape is a polygon.

       Yes, it is a polygon because it has all straight sides.

  Step 2   Count the number of sides and angles.

       There are 3 sides and 3 angles.

  Step 3   Identify the shape.

       A polygon with 3 sides and 3 angles is a triangle.

Solution   **The two-dimensional shape is a triangle.**

Look at the three angles in the triangle in Example 1. There is 1 right angle and 2 angles that are smaller than right angles. Any triangle with a right angle is called a **right triangle**.

Example 2

What is the name of this two-dimensional shape?

Strategy   **Count the number of sides and angles.**

   Step 1   Decide if the shape is a polygon.

   Yes, it is a polygon because it has all straight sides.

   Step 2   Count the number of sides and angles.

   There are 5 sides and 5 angles.

   Step 3   Identify the shape.

   A polygon with 5 sides and 5 angles is a pentagon.

Solution   **The two-dimensional shape is a pentagon.**

Example 3

Describe the sides and angles of the pentagon in Example 2.

Strategy   **Identify the types of angles and sides.**

   Step 1   Describe the types of angles.

   The angles on the left side of the pentagon form square corners.

   So, there are 2 right angles.

   There are also 2 obtuse angles.

   The angle on the right side is an acute angle.

Step 2   Describe the sides.

The top and bottom sides are parallel.

The left side meets the top and bottom sides at right angles.
So, there are 2 pairs of perpendicular sides.

The 2 right sides of the pentagon intersect.

They form 2 obtuse angles and 1 acute angle.

Solution   **The pentagon has 2 right angles, 2 obtuse angles, and 1 acute angle. It has parallel, perpendicular, and intersecting sides.**

Quadrilaterals have 4 sides and 4 angles. They are classified by the lengths of their sides and the types of angles. Here are some quadrilaterals you should know.

Name	Diagram	Properties
Parallelogram		It has two pairs of parallel sides. The opposite sides are equal.
Rhombus		It is a parallelogram with 4 equal sides.
Rectangle		It is a parallelogram with 4 right angles.
Square		It is a rectangle with 4 equal sides.
Trapezoid		It has exactly 1 pair of parallel sides.

Example 4

What is the name of this polygon? Be as specific as possible.

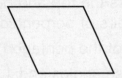

Strategy   **Look at the sides and angles.**

Step 1   Decide if the polygon is a quadrilateral.

The polygon has 4 straight sides.

It is a quadrilateral.

Step 2   Decide if the polygon is a parallelogram.

The polygon has 2 pairs of parallel sides.

It is a parallelogram.

Step 3   Look at the angles.

The polygon does not have any right angles.

It has 2 angles that are smaller than right angles.

It also has 2 angles that are greater than right angles.

So, there are 2 acute angles and 2 obtuse angles.

Step 4   Decide if the sides are the same length.

The polygon appears to have 4 equal sides.

Step 5   Name the polygon.

A rhombus is a quadrilateral with 4 sides that are the same length and 2 pairs of parallel sides.

Solution   **The polygon is a rhombus.**

Coached Example

What is the name of this two-dimensional shape? Be as specific as possible.

Decide if the shape is a polygon.

Does the shape have straight sides? _____

Is the shape a polygon? _____

Count the number of sides and angles.

How many straight sides does the shape have? _____

How many angles does the shape have? _____

Is the shape a quadrilateral? _____

Does the shape have any right angles? _____

Is the shape a rectangle? _____

What types of angles does the shape have?

_____ angles and _____ angles

Does the shape have parallel sides? _____

How many pairs of parallel sides does the shape have? _____

Which quadrilateral has only 1 pair of parallel sides? _____

The name of this two-dimensional shape is _____.

Lesson Practice

Choose the correct answer.

1. Which is **not** a polygon?

   A.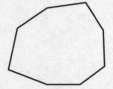

   B.

   C.

   D.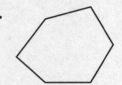

2. Which best describes this shape?

   A. quadrilateral
   B. square
   C. rhombus
   D. hexagon

3. Which of these shapes has right angles?

   A.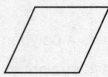

   B.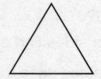

   C.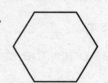

   D.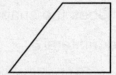

4. Wendy was driving when she passed a yield sign.

   Which best describes the angles in the sign?

   A. 3 right angles
   B. 3 acute angles
   C. 3 obtuse angles
   D. 2 acute angles and 1 right angle

5. How many pairs of parallel lines are in this shape?

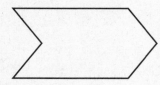

   A. 1

   B. 3

   C. 4

   D. 5

6. How many angles does an octagon have?

   A. 6

   B. 7

   C. 8

   D. 10

7. Which appears to be a right triangle?

   A.

   B.

   C.

   D.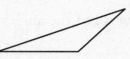

8. Which is the name of this two-dimensional shape?

   A. octagon

   B. hexagon

   C. pentagon

   D. trapezoid

9. Jessica drew the shape below.

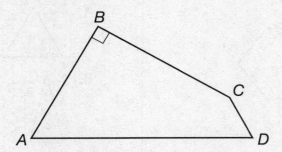

A. Describe all of the angles in the shape.

_____

_____

B. Does the shape have any parallel or perpendicular sides? Explain your answer.

_____

_____

_____

C. Name the shape that Jessica drew.

_____

Symmetry

Common Core State Standard:
4.G.3

Getting the Idea

A figure with **line symmetry** can be divided into two matching parts.

The line that divides the figure is a **line of symmetry**. Some figures have 1 line of symmetry and others have more than 1 line of symmetry.

The equilateral triangle below has 3 lines of symmetry.

Lines of Symmetry

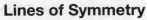

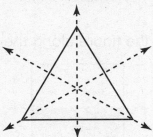

Example 1

Which figure(s) shows a line of symmetry?

| Figure A | Figure B | Figure C | Figure D |

Strategy   **Imagine folding each figure along the line so that the parts match exactly.**

Step 1   Look at Choice A.

Imagine folding the figure along the line.
The parts do not match.

Choice A does not show a line of symmetry.

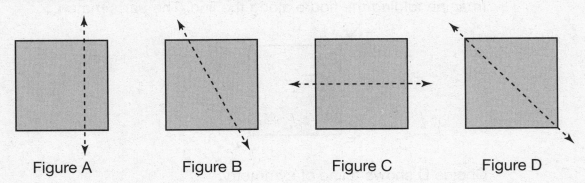

Step 2   Look at Choice B.

Imagine folding the figure along the line. The parts do not match.

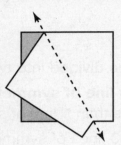

Choice B does not show a line of symmetry.

Step 3   Look at Choice C.

Imagine folding the figure along the line. The parts match.

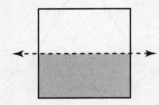

Choice C shows a line of symmetry.

Step 4   Look at Choice D.

Imagine folding the figure along the line. The parts match.

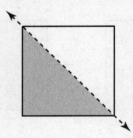

Choice D shows a line of symmetry.

Solution   **Figures C and D show a line of symmetry.**

Example 2

Which figure has line symmetry?

Strategy	**Draw lines on each figure.**
	Imagine folding on the lines to see if the parts match.

Step 1   Draw different lines on the first figure.

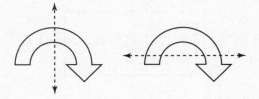

The parts do not match.

The figure does not have line symmetry.

Step 2   Draw different lines on the second figure.

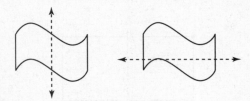

The parts do not match.

The figure does not have line symmetry.

Step 3   Draw different lines on the third figure.

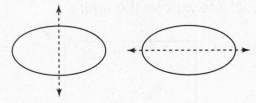

The parts match.

Both dashed lines are lines of symmetry.

The figure has line symmetry.

Solution   **The third figure has line symmetry.**

Example 3

Does this figure have line symmetry? If so, how many lines of symmetry does it have?

Strategy   **Test different lines on the rectangle.**
Imagine folding on the lines to see if the parts match.

Step 1   Draw a horizontal line through the middle of the figure.

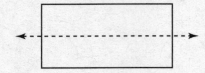

The parts match.

The dashed line is a line of symmetry.

The figure has line symmetry.

Step 2   Draw a vertical line through the middle of the figure.

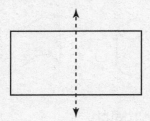

The parts match.

The dashed line is a line of symmetry.

Step 3   Draw a diagonal line across the figure.

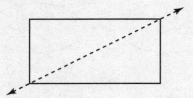

When folded across the line, the parts do not match.

The dashed line is not a line of symmetry.

Solution   **The rectangle has 2 lines of symmetry.**

Coached Example

Which block letters have line symmetry?

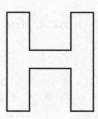

E S H

Look at the letter E.

Draw lines to see if it has line symmetry.

It has _____ line of symmetry.

Does the letter E have line symmetry? _____

Look at the letter S.

Draw lines to see if it has line symmetry.

It has _____ lines of symmetry.

Does the letter S have line symmetry? _____

Look at the letter H.

Draw lines to see if it has line symmetry.

It has _____ lines of symmetry.

Does the letter H have line symmetry? _____

The letters _____ and _____ have line symmetry.

Lesson Practice

Choose the correct answer.

1. Which figure shows a line of symmetry?

   A.

   B.

   C.

   D.

2. Which triangle has only 1 line of symmetry?

   A.

   B.

   C.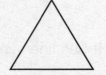

   D.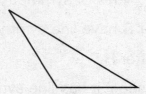

3. Which has exactly 2 lines of symmetry?

   A.

   B.

   C.

   D. R

4. How many lines of symmetry does this figure have?

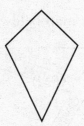

A. 0   **C.** 2

B. 1   **D.** 4

5. How many lines of symmetry does this figure have?

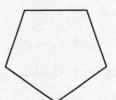

A. 3   **C.** 5

B. 4   **D.** 7

6. Which figure has the most lines of symmetry?

A.   **C.**

B. 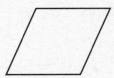   **D.**

7. How many lines of symmetry does this block letter have?

A. 0

B. 1

C. 2

D. 3

8. Trevon drew the figure below.

A. Draw a line of symmetry on the figure.

B. Does the figure have line symmetry? Explain how you know.

_____

_____

_____

Domain 5: Cumulative Assessment for Lessons 37–39

1. Which shows perpendicular lines?

   A.

   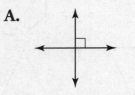

   B.

   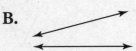

   C.

   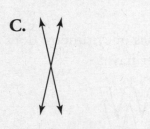

   D.

2. What type of angle is shown below?

   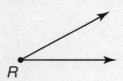

   A. It is an obtuse angle.

   B. It is a straight angle.

   C. It is an acute angle.

   D. It is a right angle.

3. Look at the map below.

   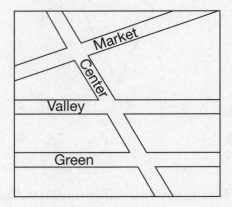

   Which streets appear to be parallel to each other?

   A. Green and Valley

   B. Valley and Market

   C. Market and Center

   D. Center and Green

4. Which is **not** true about this shape?

   A. It is a quadrilateral.

   B. The opposite sides are parallel.

   C. Two angles are greater than right angles.

   D. It has perpendicular line segments.

5. Nicole drew a polygon with 2 pairs of parallel sides and 4 right angles. The length is twice as long as the width. What shape best describes the polygon Nicole drew?

A. rectangle

B. square

C. rhombus

D. trapezoid

6. Which shape does **not** have any parallel sides?

A.

B.

C.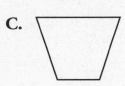

D.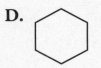

7. Which is a right triangle?

A.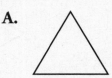

B.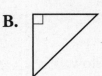

C.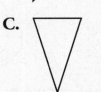

D.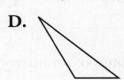

8. Which figure has the most lines of symmetry?

A.

B.

C.

D.

9. How many lines of symmetry does this figure have?

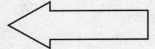

_____

10. Look at the figure below.

   A. How many lines of symmetry does the figure have?

_____

   B. Draw all the lines of symmetry on the figure below.

Glossary

acute angle an angle with a measure less than 90 degrees (Lesson 37)

add (addition) to find the total when two or more groups are joined (Lesson 12)

addend a number to be added (Lesson 12)

angle a figure formed by two rays that have the same endpoint (Lesson 35)

area the number of square units needed to cover a figure (Lesson 34)

array an arrangement of objects in equal rows and columns (Lesson 3)

associative property of multiplication a property that states that when you multiply, the grouping of the factors does not change the product (Lesson 5)

base-ten numeral a number written using only its digits (Lesson 1)

benchmark a common number that can be compared to another number (Lesson 20)

capacity the measure of how much liquid a container holds; also called liquid volume (Lesson 31)

centimeter (cm) a metric unit of length; 1 centimeter = 10 millimeters (Lesson 32)

circle a two-dimensional shape in which all points are an equal distance from the center (Lesson 38)

common denominator the same denominator in two or more fractions (Lesson 20)

commutative property of multiplication a property that states that when you multiply, the order of the factors does not change the product (Lesson 5)

compatible numbers numbers that are close in value to the exact numbers and that are easy to compute with (Lesson 16)

composite number a whole number that has more than one factor pair (Lesson 11)

cup a customary unit of capacity; 2 cups = 1 pint (Lesson 31)

day (d) a unit of time; 1 day = 24 hours (Lesson 29)

decimal a number with a decimal point (Lesson 25)

decimal point (.) a period separating the ones from the tenths in a decimal (Lesson 25)

degree (°) a unit for measuring angles (Lesson 35)

denominator the bottom number in a fraction, which tells how many equal parts in the whole or group (Lesson 18)

difference the answer in a subtraction problem (Lesson 13)

distributive property of multiplication a property that states that when you multiply a number by a sum, you can multiply the number by each addend of the sum and then add the products (Lesson 6)

divide (division) to find the number of equal groups or the number in each group (Lesson 7)

dividend the number to be divided (Lesson 7)

divisor the number by which the dividend is divided (Lesson 7)

elapsed time the amount of time that has passed from a beginning time to an end time (Lesson 29)

endpoint a point on the end of a line segment or ray (Lesson 35)

equivalent fractions two or more fractions that name the same value, but have different numerators and denominators (Lesson 18)

estimate a number that is close to the exact amount (Lesson 15)

expanded form a way of writing a number that shows the sum of the values of each digit (Lesson 1)

fact family a group of related addition and subtraction facts or multiplication and division facts that use the same numbers (Lesson 7)

factor a number that is multiplied to get a product (Lessons 3, 11)

foot (ft) a customary unit of length; 1 foot = 12 inches (Lesson 32)

fraction a number that names part of a whole or part of a group (Lesson 18)

gallon a customary unit of capacity; 1 gallon = 4 quarts (Lesson 31)

gram (g) a metric unit of mass; 1,000 grams = 1 kilogram (Lesson 30)

greatest common factor (GCF) the greatest factor that is common to two or more numbers (Lesson 18)

hexagon a two-dimensional shape with 6 sides and 6 angles (Lesson 38)

hour (hr) a unit of time; 1 hour = 60 minutes (Lesson 29)

improper fraction a fraction with a numerator that is equal to or greater than its denominator (Lesson 19)

inch (in.) a customary unit of length; 12 inches = 1 foot (Lesson 32)

intersecting lines lines that cross (Lesson 37)

inverse operations operations that undo each other, such as addition and subtraction, and multiplication and division (Lessons 7, 8, 13)

is equal to (=) a symbol that shows that two quantities have the same value (Lesson 2)

is greater than (>) a symbol that shows that the first quantity is greater than the second quantity (Lesson 2)

is less than (<) a symbol that shows that the first quantity is less than the second quantity (Lesson 2)

kilogram (kg) a metric unit of mass; 1 kilogram = 1,000 grams (Lesson 30)

kilometer (km) a metric unit of length; 1 kilometer = 1,000 meters (Lesson 32)

length the measure of how long, wide, or tall something is (Lesson 32)

like denominators two or more denominators that are the same (Lesson 21)

line a straight path that goes in two directions without end (Lesson 37)

line of symmetry a line that divides a figure in which the halves match exactly (Lesson 39)

line plot a graph that uses dots above a number line to record data (Lesson 36)

line segment a part of a line that has two endpoints (Lesson 37)

line symmetry a property of a figure in which a figure can be divided into 2 matching halves (Lesson 39)

liquid volume the measure of how much liquid a container holds; also called capacity (Lesson 31)

liter a metric unit of capacity; 1 liter = 1,000 milliliters (Lesson 31)

mass the amount of matter in an object (Lesson 30)

meter (m) a metric unit of length; 1 meter = 100 centimeters (Lesson 32)

mile (mi) a customary unit of length; 1 mile = 1,760 yards (Lesson 32)

milliliter (mL) a metric unit of capacity; 1,000 milliliters = 1 liter (Lesson 31)

millimeter (mm) a metric unit of length; 10 millimeters = 1 centimeter (Lesson 32)

minuend the number being subtracted from in a subtraction problem (Lesson 13)

minute (min) a unit of time; 60 minutes = 1 hour (Lesson 29)

mixed number a number with a whole number part and a fraction part (Lesson 19)

month (mo) a unit of time; 12 months = 1 year (Lesson 29)

multiple the product of a number and another number (Lessons 10, 11)

multiplicative identity property of 1 a property that states that when you multiply a number by 1, the product is that number (Lesson 5)

multiply (multiplication) to find a total when there are equal groups; a shortcut for repeated addition (Lesson 3)

number name a way of writing numbers using words (Lesson 1)

numerator the top number in a fraction, which tells how many equal parts are being considered (Lesson 18)

obtuse angle an angle with a measure greater than 90 degrees, but less than 180 degrees (Lesson 37)

octagon a two-dimensional shape with 8 sides and 8 angles (Lesson 38)

ounce (oz) a customary unit of weight; 16 ounces = 1 pound (Lesson 30)

parallel lines lines that remain the same distance apart and never meet (Lesson 37)

parallelogram a quadrilateral with 2 pairs of parallel sides; the opposite sides are equal (Lesson 38)

pattern a series of numbers or figures that follows a rule (Lesson 17)

pentagon a two-dimensional shape with 5 sides and 5 angles (Lesson 38)

perimeter the distance around a figure, measured in units (Lesson 33)

perpendicular lines intersecting lines that cross to form 4 right angles (Lesson 37)

pint a customary unit of capacity; 1 pint = 2 cups (Lesson 31)

place value the value of a digit based on its position in a number (Lesson 1)

point a particular place or location (Lesson 37)

polygon a type of two-dimensional shape with straight sides (Lesson 38)

pound a customary unit of weight; 1 pound = 16 ounces (Lesson 30)

prime number a whole number that has exactly one factor pair, 1 and itself (Lesson 11)

product the answer in a multiplication problem (Lesson 3)

protractor a tool used to measure angles (Lesson 35)

quadrilateral a two-dimensional shape with 4 sides and 4 angles (Lesson 38)

quart a customary unit of capacity; 1 quart = 2 pints (Lesson 31)

quotient the answer in a division problem (Lesson 7)

ray a part of a line that has an endpoint at one end and goes on forever in the other direction (Lessons 35, 37)

rectangle a parallelogram with 4 right angles (Lesson 38)

regroup to rename a number, such as 10 tens as 1 hundred (Lesson 12)

remainder a number that is left after division has been completed (Lesson 9)

rhombus a parallelogram with 4 equal sides (Lesson 38)

right angle an angle that measures exactly 90° (Lesson 37)

right triangle a triangle with one right angle (Lesson 38)

round to replace a number with one that tells about how much or about how many (Lesson 14)

rule a description of how the terms are related in a pattern (Lesson 17)

simplest form a form of a fraction whose numerator and denominator have only 1 as a common factor (Lesson 18)

square a rectangle with 4 equal sides (Lesson 38)

subtract (subtraction) to find how many are left when a quantity is taken away (Lesson 13)

subtrahend the number that is subtracted in a subtraction problem (Lesson 13)

sum the answer in an addition problem (Lesson 12)

term a number in a number pattern (Lesson 17)

trapezoid a quadrilateral with exactly 1 pair of parallel sides (Lesson 38)

triangle a two-dimensional shape with 3 sides and 3 angles (Lesson 38)

two-dimensional shape a flat figure (Lesson 38)

variable a letter or symbol used to represent a value that is unknown (Lesson 3)

vertex the common endpoint where two rays meet in an angle (Lesson 35)

week (wk) a unit of time; 1 week = 7 days (Lesson 29)

weight a measurement that tells how heavy an object is (Lesson 30)

whole number any of the numbers 1, 2, 3, and so on (Lesson 1)

yard (yd) a customary unit of length; 1 yard = 3 feet (Lesson 32)

year (yr) a unit of time; 1 year = 12 months (Lesson 29)

**Crosswalk Coach for the Common Core State Standards
Mathematics, Grade 4**

Summative Assessment:
Domains 1–5

Name: _____

Session 1

1. In 2008, the population of Denham Springs was ten thousand, three hundred eight. What is the population of Denham Springs written in standard form?

    A. 10,038

    B. 10,308

    C. 10,318

    D. 13,008

2. Natalie ate $\frac{3}{8}$ of an apple. Later, she ate another $\frac{3}{8}$ of the apple. What fraction of the apple did she eat in all?

    A. $\frac{2}{8}$ **C.** $\frac{6}{16}$

    B. $\frac{6}{8}$ **D.** $\frac{8}{8}$

3. Whitney drank $\frac{1}{5}$ cup of milk before lunch. She drank $\frac{3}{5}$ cup of milk with her lunch. Which number sentence could be used to find how much milk she drank?

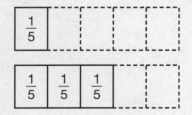

    A. $\frac{2}{5} = \frac{3}{5} - \frac{1}{5}$

    B. $\frac{3}{5} = \frac{1}{5} + \frac{2}{5}$

    C. $\frac{4}{5} = \frac{1}{5} + \frac{3}{5}$

    D. $\frac{4}{5} = \frac{2}{5} + \frac{2}{5}$

4. Maple Avenue is 0.6 kilometer long.

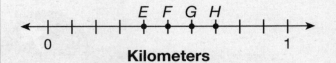

Which point on the number line represents 0.6 kilometers?

    A. point E

    B. point F

    C. point G

    D. point H

5. Which shape has parallel sides?

    A.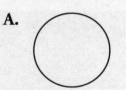

    B.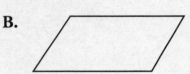

    C.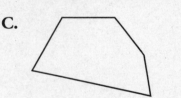

    D.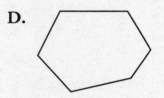

6. The grid has 0.46 shaded.

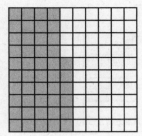

Which sum of fractions shows 0.46?

A. $\frac{4}{10} + \frac{6}{100}$

B. $\frac{4}{10} + \frac{6}{10}$

C. $\frac{4}{100} + \frac{6}{100}$

D. $\frac{4}{100} + \frac{6}{10}$

7. Deon earned $2,416 each week for 8 weeks. How much money did he earn in all?

A. $11,294

B. $19,288

C. $19,328

D. $40,328

8. Carol pitched $\frac{3}{6}$ of the softball game. Josie pitched $\frac{2}{6}$ of the game. What fraction of the softball game did Carol and Josie pitch combined?

A. $\frac{1}{12}$

B. $\frac{5}{12}$

C. $\frac{2}{3}$

D. $\frac{5}{6}$

9. What is the measure of the angle below? Use a protractor.

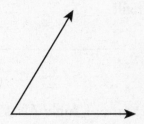

A. 45°

B. 60°

C. 120°

D. 135°

10. Which fraction is equal to 0.7?

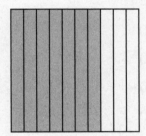

A. $\frac{1}{7}$ **C.** $\frac{7}{100}$

B. $\frac{7}{10}$ **D.** $\frac{70}{10}$

11. Los Angeles is 2,108 miles from Louisville. Boston is 941 miles from Louisville. How many miles closer to Louisville is Boston than Los Angeles?

A. 1,157 miles

B. 1,167 miles

C. 1,257 miles

D. 3,049 miles

12. If 7 people each eat $\frac{3}{8}$ pound of potato salad, how much potato salad will be eaten in all?

_____

13. The measure of angle I is 130°. A part of angle I measures 70°.

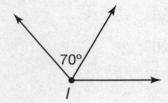

What is the measure of the other part of angle I?

A. 130°

B. 90°

C. 70°

D. 60°

14. What is the area of this rectangle?

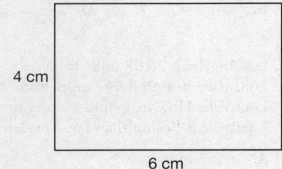

4 cm

6 cm

A. 10 square centimeters

B. 20 square centimeters

C. 24 square centimeters

D. 48 square centimeters

15. Which will have the same product as $4 \times \frac{3}{4}$?

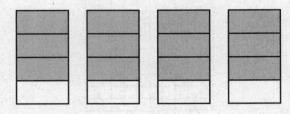

A. $12 \times \frac{1}{3}$

B. $12 \times \frac{1}{2}$

C. $12 \times \frac{1}{4}$

D. $12 \times \frac{1}{5}$

16. Which number has 6 and 7 as factors?

A. 14   **C.** 35

B. 24   **D.** 42

17. Which figure has line symmetry?

A.

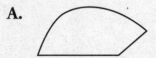

B.

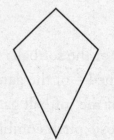

C.

D.

18. Which is a prime number?

A. 21

B. 23

C. 25

D. 27

19. Diana ran 0.8 mile in 5 minutes. Which is 0.8 written as a fraction?

A. $\frac{1}{8}$

B. $\frac{8}{100}$

C. $\frac{80}{10}$

D. $\frac{8}{10}$

20. Susan ran 5 laps that are each $\frac{8}{10}$ mile. How many miles did she run in all?

A. 4 miles

B. 5 miles

C. 8 miles

D. 40 miles

21. How many times greater is the 7 in the ten thousands place than the 7 in the hundreds place in this number?

73,746

A. 1

B. 10

C. 100

D. 1,000

22. The black shoes cost $54. The blue shoes cost 4 times as much. How much do the blue shoes cost?

A. $206

B. $210

C. $216

D. $218

23. Malik bought a notebook for $3.29. He paid with a $5 bill. How much change should Malik receive from the cashier?

_____

24. Kendra scored 2,355 points playing a computer game. The next time she played, Kendra scored 2,870 points. How many points did Kendra score in all?

A. 4,125

B. 5,125

C. 5,135

D. 5,225

25. Which appears to be a right triangle?

A.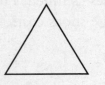

B.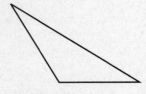

C.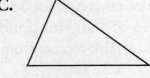

D.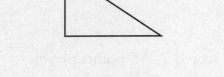

26. Josh has 3 trophies. He plans to have 7 times as many trophies when he grows up. Which number sentence could be used to find how many trophies in all?

A. $28 = 7 \times 3 + 7$

B. $28 = 3 + 7 + 3 + 7 + 8$

C. $21 = 3 \times 7$

D. $21 = 7 + 7 + 3 + 4$

27. Which lists the decimals from greatest to least?

A. 0.36   0.42   0.4

B. 0.4   0.36   0.42

C. 0.42   0.4   0.36

D. 0.42   0.36   0.4

28. Which number sentence is true?

A. $0.36 > 0.36$

B. $0.37 > 0.27$

C. $0.2 = 0.45$

D. $0.5 < 0.35$

29. Annie's patio is in the shape of a rectangle. It has an area of 150 square feet. The length is 15 feet. What is the width of the patio?

A. 8 feet

B. 9 feet

C. 10 feet

D. 20 feet

30. Abram has 15 kilograms of sand. If he makes 3 piles of sand, how many grams of sand will be in each pile?

A. 5 grams

B. 15 grams

C. 500 grams

D. 5,000 grams

31. What is $5\frac{3}{8} - 2\frac{1}{8}$?

    A. 3

    B. $3\frac{1}{8}$

    C. $3\frac{2}{8}$

    D. $3\frac{3}{8}$

32. Yesterday, Nikki bought $\frac{3}{8}$ pound of cheddar cheese and $\frac{5}{8}$ pound of swiss cheese. How much cheese did she buy?

    A. $\frac{2}{8}$ pound

    B. $\frac{8}{16}$ pound

    C. $\frac{15}{8}$ pound

    D. 1 pound

33. Which number sentence is true?

    A. 36,427 > 36,274

    B. 42,394 = 42,934

    C. 51,792 > 52,630

    D. 57,531 < 57,315

34. Draw a line of symmetry on the figure below.

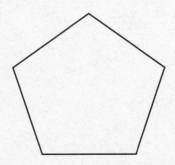

35. Which table shows the relationship between years and months?

A.

Year	Months
1	7
2	14
3	21
4	28

B.

Year	Months
1	6
2	12
3	18
4	24

C.

Year	Months
1	12
2	24
3	36
4	48

D.

Year	Months
1	60
2	120
3	180
4	240

36. Which measure is equal to 1 meter?

    A. 10 millimeters

    B. 100 millimeters

    C. 10 centimeters

    D. 100 centimeters

37. Which number sentence is true?

A. $2.83 > 2.8$

B. $3.1 < 3.07$

C. $4.05 > 4.2$

D. $5.3 < 5.24$

38. A wooden board is 56 inches long before it is cut into short equal pieces. That is 7 times as long as each short piece. How long is each short piece?

A. 6 inches

B. 7 inches

C. 8 inches

D. 9 inches

39. What is true about this shape?

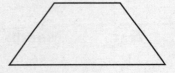

A. It has one pair of parallel sides.

B. It has two pairs of parallel sides.

C. It has one pair of perpendicular sides.

D. It has parallel and perpendicular sides.

40. Nancy measured the length of a paper clip as $\frac{7}{8}$ inch. Fred measured the paper clip as $\frac{5}{8}$ inch. What is the difference in the measurements?

A. $\frac{1}{8}$ inch

B. $\frac{2}{8}$ inch

C. $\frac{3}{8}$ inch

D. $1\frac{4}{8}$ inches

41. Which symbol makes this sentence true?

$$\frac{1}{4} \bigcirc \frac{1}{5}$$

A. $>$

B. $<$

C. $=$

D. $+$

42. The rectangle below is $\frac{5}{10}$ shaded.

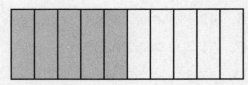

Which is an equivalent fraction to $\frac{5}{10}$?

A. $\frac{1}{10}$

B. $\frac{1}{5}$

C. $\frac{1}{2}$

D. $\frac{15}{100}$

43. Mrs. Dante gave her 4 grandchildren $80 to go to the movies. The grandchildren were allowed to keep and equally share any leftover money after paying for the tickets. The tickets cost $36 in all. How much money did each grandchild get?

   A. $8

   B. $9

   C. $10

   D. $11

44. The width of a rectangle is 5 inches. The length of the rectangle is 6 inches. What is the area of the rectangle?

   A. 30 square inches

   B. 25 square inches

   C. 24 square inches

   D. 11 square inches

45. Name all of the factor pairs for the number 64.

   _____

   _____

46. Which shows 84,267 rounded to the nearest thousand?

   A. 80,000

   B. 84,000

   C. 84,300

   D. 85,000

47. What fraction of a circle is a 1-degree angle?

   A. $\frac{1}{360}$

   B. $\frac{10}{360}$

   C. $\frac{45}{360}$

   D. $\frac{90}{360}$

48. The ground floor of a theater has 36 rows. Each row has 42 seats. How many seats does the ground floor of the theater have?

   A. 1,402

   B. 1,502

   C. 1,512

   D. 1,612

Session 2

49. Lori is making costumes. She needs $2\frac{3}{4}$ yards of red fabric and $1\frac{2}{4}$ yard of green fabric.

A. How much more red fabric than green fabric does she need? Show your work.

_____

_____

_____

B. How many yards of fabric does she need in all? Show your work. Write your answer in simplest form.

_____

_____

_____

50. The Apple family has a collection of 1,732 DVDs. The DVDs will be stored equally on 8 bookshelves. Any extra DVDs will be put in a basket next to the sofa.

A. How many DVDs will be on each bookshelf? Show your work.

B. How many DVDs will be in the basket? Explain your answer.

_____

_____

_____

51. Derek asked some classmates how many hours they spent doing their math homework. The results are shown below.

$\frac{2}{4}$ hour	$\frac{1}{4}$ hour	1 hour	$\frac{2}{4}$ hour	$\frac{3}{4}$ hour
$\frac{1}{4}$ hour	0 hours	$\frac{2}{4}$ hour	$\frac{2}{4}$ hour	$\frac{1}{4}$ hour
1 hour	$\frac{2}{4}$ hour	$\frac{3}{4}$ hour	$\frac{1}{4}$ hour	$\frac{3}{4}$ hour

A. Make a line plot of Derek's results. Be sure to include a title and label the number line.

B. How many more students spent $\frac{1}{2}$ hour than 1 hour doing their math homework? Explain how you found your answer.

_____

_____

_____

_____

Summative Assessment

Math Tools: Place-Value Models

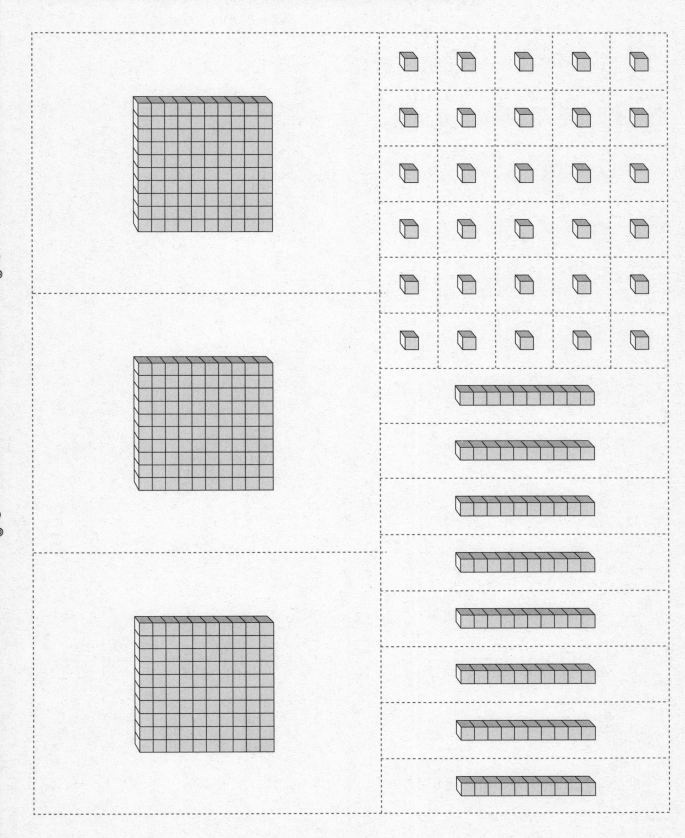

Math Tools: Fraction Circles

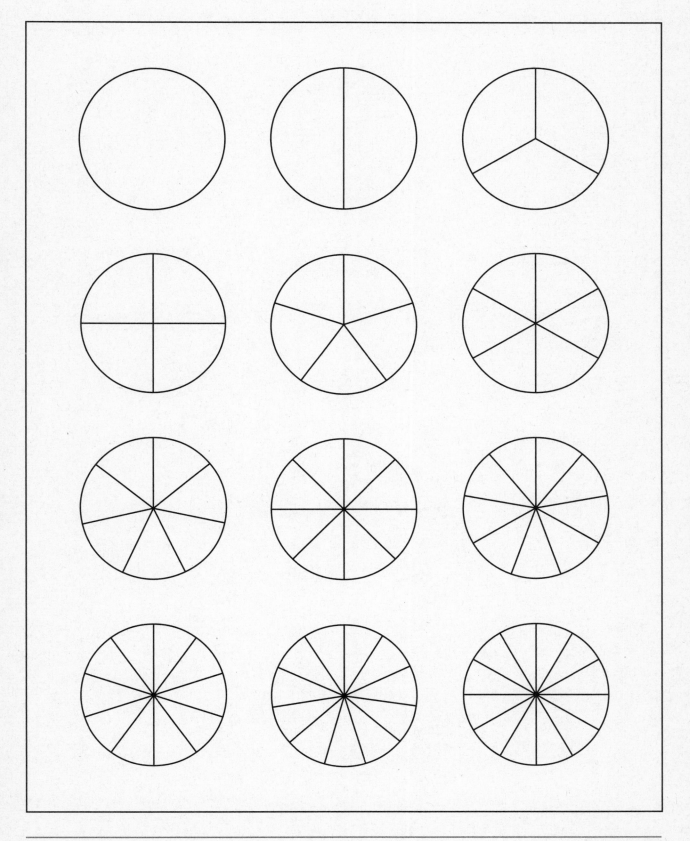

Math Tools: Fraction Strips

NOTES

NOTES

NOTES

NOTES

NOTES